PROBABILITY: ELEMENTS OF THE MATHEMATICAL THEORY

PROBABILITY

ELEMENTS OF THE MATHEMATICAL THEORY

C. R. Heathcote

Reader in Statistics, Australian National University, Canberra

WILEY INTERSCIENCE DIVISION
JOHN WILEY & SONS, INC., NEW YORK

FIRST PUBLISHED IN 1971

ISBN 0 04 519005 4 *Cased*
0 04 519006 2 *Paper*

Library of Congress Catalog Card No: 70-142998

PRINTED IN GREAT BRITAIN
in 10 point Times Roman type
BY PAGE BROS. (NORWICH) LTD., NORWICH

CONTENTS

Sections marked * can be omitted at a first reading

PREFACE

Several excellent books on probability have appeared in recent years but most are for readers with some degree of mathematical sophistication and experience. The present text is meant for undergraduates beginning the study of mathematical statistics and probability after completing a full first year course on calculus and real variables. Prerequisites are some set theory, Riemann integration, and familiarity with the elementary operations of analysis. Emphasis is placed as much on method as on basic theory, and, whilst it is hoped that the book will be of some general interest, the major aim is to provide what in the author's opinion is a suitable preparation for further study in the field.

Material for a course of approximately thirty six lectures is contained in the unstarred sections of Chapters 1, 3, 4, and 5. Parts of Chapter 5 will be recognised as measure theory in a special setting but neither the term nor the general theory are invoked. However, a few important and essential theorems from analysis and measure theory have been stated without proof but this seems justified on the grounds that the discussion would perforce be relatively trivial without them. An attempt has always been made to motivate these results and to point out in what way the proofs require techniques beyond the scope of an introductory treatment.

Starred sections develop particular facets of the theory, §3.4 and §4.4 being included for intending statisticians whereas the remainder are for probabilists and mathematics majors. In particular §§3.5, 4.5, and 5.5 have been used for a lecture series on Markov chains designed as a preliminary to a more systematic treatment of stochastic processes.

It is suggested that only the main results of Chapter 2 be noted at a first reading and the chapter used mainly for reference as required. The space allotted to generating functions and standard theorems such as those of Taylor and Abel may seem anomalous. However the author's experience has been that the present generation of mathematics undergraduates, whilst on the whole coping confidently with abstract arguments, frequently experience some difficulty when working with certain concepts of classical real variable theory (typically questions related to series and the convergence of integrals and sums). Partly for this reason, but mainly because it seemed useful to bring together in probabilistic terminology several of the main results concerning an important tool, the treatment of generating functions in the first three sections is of some length.

Introductory courses for statisticians tend to take (wrongly in the author's view) many technical matters for granted and readers in this category may find much of Chapters 2 and 5, in particular, neither necessary nor interesting. On the other hand probabilists have traditionally paid attention as much to analytical detail as to general principles and the abstract aspect of the subject. Indeed the purpose of this book will be amply served if it convinces students early in their career that the study of probability requires both the generality given by theories of measures and integration as well as the hard labour and technique necessary to solve special problems.

References have been kept to a minimum, as is appropriate in an introductory exposition. The books

W. Feller, *An Introduction to Probability Theory and its Applications,* Vol I (3rd ed.), Wiley International Edition, New York 1968, and Vol II, Wiley, New York, 1966.

E. Parzen, *Modern Probability Theory and its Applications,* Wiley, New York, 1960,

are referred to in the text by author's name only as Feller I or II, and Parzen, respectively. The most frequently cited is Feller I, whose influence on the present book will be apparent.

Finally, thanks are due to colleagues who read various chapters; to Mrs. J. Radley, Mrs. I. Kewley, and Miss C. Martin for expertly typing several drafts of the manuscript; to the publishers for their help and courtesy; and mainly to those students of the Australian National University and London School of Economics who, as captive audiences, had no choice but to be introduced to probability in the manner presented here.

Principal Notations

$\Omega = \{\omega\}$	sample space	*page* 1
$A, B, C, \ldots,$	events on a sample space	2
$\bar{A}$	complement of A with respect to the sample space	2
$\emptyset$	null event	3
$\mathscr{F}$	field or σ field of events on Ω	3
P	probability function	3
$(\Omega, \mathscr{F}, P)$	probability space	3
$X, Y, \ldots,$	random variables	26
$X \frown N(\mu, \sigma^2)$	X is normally distributed with parameters μ, σ^2 (in general $X \frown \mathscr{L}$ means that X is distributed according to the law $\mathscr{L}$)	32
$Eg(X)$	expectation of $g(X)$	35, 37
$\mathrm{Var}\, X$	variance of X	38
$O(x), o(x)$	order notation	52
$\lim\limits_{x \to 0+}, \lim\limits_{x \to 0-}$	limits from above and below	54
$I_A(\omega)$	indicator function of A	104
$\mathrm{Cov}(X, Y)$	covariance of X and Y	115
$P(A \mid B)$	conditional probability of A given B	123
$\xrightarrow{d}$	limit in distribution	161
$\xrightarrow{P}$	limit in probability	173
$\limsup\limits_{n \to \infty}$	limit superior	215
$\liminf\limits_{n \to \infty}$	limit inferior	216
$\xrightarrow{a.s.}$	limit almost surely	222

The notation Pr(...) represents the statement 'the probability of (...)'

Chapter 1

PROBABILITY SPACES AND RANDOM VARIABLES

Probability theory is concerned with the mathematical analysis of quantities derived from observations of phenomena whose occurrence involves a chance element. The result of tossing a 'fair' coin cannot be predicted with certainty and any mathematical theory of probability must give meaning to and answer questions such as 'what is the probability of n heads in N tosses of such a coin?' Coin tossing is a simple example of the class of phenomena with which the theory is concerned.

For a mathematical theory the essential ingredients are that an act or experiment is performed (e.g. tossing a coin) all possible outcomes of which can at least in principle be specified and are observable (heads or tails), and that a rule is given whereby for each possible outcome or set of outcomes there can be defined in a consistent way a number called the probability of the outcome or set of outcomes in question.

Given a chance phenomenon the set of possible outcomes is called the *sample space* and will be denoted by Ω. The individual outcomes, generically denoted by ω, comprise the elements of the sample space, thus $\Omega = \{\omega\}$. Our first concern will be to set up the requisite formalism for an axiomatic definition of probability. This is done in § 1.1 commencing with the simple case of a sample space containing only finitely many elements.

1.1 PROBABILITY SPACES

Suppose we are given a sample space Ω consisting of the m points $\omega_1, \omega_2, \ldots, \omega_m$;

$$\Omega = \{\omega_1, \omega_2, \ldots, \omega_m\}.$$

Example 1 One toss of a coin. The only possible outcomes are heads or tails. Let $\omega_1 = H$ denote heads and $\omega_2 = T$ denote tails. Then the sample space Ω contains two elements, $\Omega = \{H, T\}$. □

Example 2 Roll of a die. Supposing the six sides of the die are numbered 1 to 6, Ω consists of the first six positive integers. □

Example 3 Roll of two dice. An outcome ω_j is a vector (a, b) in which a exhibits the number shown by the first die and b that shown by the second. Ω contains $6^2 = 36$ elements, namely the set of vectors $\{(a, b) \mid a, b = 1, 2, 3, 4, 5, 6\}$. □

The outcomes ω_j are called *elementary events* or *sample points* and the three examples show that each ω_j may be a symbol, a number, or a vector (or indeed a much more general quantity) depending on the particular phenomenon observed. From the point of view of the general theory the ω_j are undefined objects.

It is frequently the case that an observer is just as interested in collections of the ω_j as in the elementary events themselves, and these collections of elementary events are called simply *events*. In Example 2 above with $\omega_j = j$, $j = 1, 2, \ldots, 6$, we have as an illustration that the event

$$A = \text{'the die shows an even number'}$$

is given by the subset

$$A = \{\omega_2, \omega_4, \omega_6\}.$$

The *complement* of A with respect to Ω, written $\overline{A}$, is also an event, namely,

$$\begin{aligned} \overline{A} &= \text{'the die does not show an even number'} \\ &= \{\omega_1, \omega_3, \omega_5\} = \Omega - A. \end{aligned}$$

Events $A, B, C, \ldots$, are then subsets of Ω. For an event A defined on a sample space the statement 'A occurs' means that an observation of the underlying chance phenomenon resulted in an ω such that $\omega \in A$. It is of course possible for two or more events to occur simultaneously. *Disjoint events* are those containing no sample points in common, that is, their intersection $AB = A \cap B$ is empty and they cannot occur simultaneously.

It is convenient to define the *certain* or *sure event* as Ω itself, since $\omega \in \Omega$ for all ω implies that the event Ω occurs at each observation. The *impossible* or *null event*, denoted by $\varnothing$, is defined to be the complement of Ω, namely $\varnothing = \overline{\Omega}$. The certain event contains all elementary events and the impossible event contains none.

With this definition of an event as a subset of the sample space it is possible to form the set $\mathscr{F}$ of *all* events (including the sure and null events) defined on a sample space $\Omega = \{\omega_1, \omega_2, \ldots, \omega_m\}$. The set $\mathscr{F}$ is a *field* in the sense that

(i) if $A, B \in \mathscr{F}$ then $A \cup B \in \mathscr{F}$,
(ii) if $A \in \mathscr{F}$ then $\overline{A} \in \mathscr{F}$,

i.e. $\mathscr{F}$ is closed under the formation of unions and complements.*

It remains to formally define what is meant by the probability of an event.

Definition 1.1.1 Given a finite sample space Ω with field of events $\mathscr{F}$, *the probability of the event* $A \in \mathscr{F}$, written $P(A)$, is a set function onto the closed interval $[0, 1]$ satisfying the conditions

(i) $0 \leqslant P(A) \leqslant 1$, all $A \in \mathscr{F}$,
(ii) $P(\Omega) = 1$,
(iii) $P(A \cup B) = P(A) + P(B)$ if $A \cap B = \varnothing$. □

This axiomatic definition of probability implies that it is an additive non-negative set function of a particular kind. Our concern is with probability defined in this way and the theory does not encompass situations in which the term may have valid everyday usage but which lack the definition of a field of events $\mathscr{F}$ over a sample space Ω.

The triple $(\Omega, \mathscr{F}, P)$ is called a *probability space* and is the point of departure in the modern treatment of probability.

Definition 1.1.1 of probability is satisfactory for finite sample spaces, However it is frequently the case that we wish to consider probability spaces in which Ω contains infinitely many points. The definition of $\mathscr{F}$ and P must be extended to cope with the circumstance of infinitely many subsets (events) of Ω. We require that the field $\mathscr{F}$ in fact be a σ *field*, namely that $\mathscr{F}$ is closed under the

* A field $\mathscr{F}$ is also closed under the formation of intersections (see Exercise 1.2.8).

formation of complements and the union of *countably* many events.* Of course $\mathscr{F}$ still contains Ω and $\varnothing$. Furthermore the set function P must now be *countably additive* so that part (iii) of Definition 1.1.1 becomes

(iii)′ If $A_1, A_2, \ldots$ are disjoint members of the σ field $\mathscr{F}$ then

$$P\left(\bigcup_{j=1}^{\infty} A_j\right) = \sum_{j=1}^{\infty} P(A_j).$$

Subject to these modifications our definition of probability continues to hold whether Ω is finite or not and in future we will always assume the existence of an appropriate probability space $(\Omega, \mathscr{F}, P)$.

1.2 PROPERTIES OF PROBABILITY SPACES

Several properties of the function P follow immediately from the definition. Some useful ones are:

(a) If $A_1, A_2, \ldots, A_n$ are disjoint then

$$P\left(\bigcup_{j=1}^{n} A_j\right) = \sum_{j=1}^{n} P(A_j) \tag{1.2.1}$$

(a direct consequence of (iii) of Definition 1.1.1 or (iii)′).

(b) A and $\bar{A}$ are disjoint and therefore

$$1 = P(\Omega) = P(A \cup \bar{A}) = P(A) + P(\bar{A})$$
$$P(\bar{A}) = 1 - P(A). \tag{1.2.2}$$

(c) Taking $A = \Omega$ in (1.2.2),

$$P(\varnothing) = 0. \tag{1.2.3}$$

(d) If $A \subseteq B$ then

$$P(A) \leqslant P(B), \tag{1.2.4}$$

for $A \subseteq B$ implies that $B = A \cup \bar{A}B$ is a decomposition of B into disjoint subsets, hence $P(B) = P(A) + P(\bar{A}B) \geqslant P(A)$.

(e) For $A, B \in \mathscr{F}$,

$$P(A \cup B) = P(A) + P(B) - P(AB). \tag{1.2.5}$$

* As in the finite case it can be shown that a σ field is also closed under the formation of countably many intersections.

The proof is again by decomposition into disjoint subsets; thus

$$A \cup B = A \cup (B - AB), \qquad B = AB \cup (B - AB),$$

and hence

$$P(A \cup B) = P(A) + P(B - AB) = P(A) + P(B) - P(AB).$$

Examples of probability spaces follow.

Example 1 Toss of a 'fair' coin. As noted before the sample space is $\Omega = \{H, T\}$ where H denotes heads and T tails. The null event $\varnothing$ contains neither H nor T—presumably it includes the possibility of the coin landing and remaining on its edge. The field $\mathscr{F}$ contains the four elements $\varnothing$, Ω, $\{H\}$ and $\{T\}$. Turning to the probability function P the assumption that the coin is 'fair' translated into our terminology implies that

$$P(\{H\}) = P(\{T\}),$$

i.e. the probability of a toss resulting in heads is the same as the probability of obtaining tails. As a consequence

$$1 = P(\Omega) = P(\{H\} \cup \{T\}) = P(\{H\}) + P(\{T\}) = 2P(\{H\})$$

and for a fair coin $P(\{H\}) = P(\{T\}) = \frac{1}{2}$. If it is not given that the coin is fair we infer only that $P(\{H\}) = 1 - P(\{T\})$. □

Example 2 Roll of a die. Assuming again that the sides are labelled by the first six positive integers the sample space is $\Omega = \{1, 2, 3, 4, 5, 6\}$ and the elements of $\mathscr{F}$ are $\varnothing$, $\{1\}, \{2\}, \ldots,$ $\{6\}, \{1, 2\}, \ldots, \Omega$. $\mathscr{F}$ contains the $2^6 = 64$ subsets of Ω, and the probability function P must be such that the probability of each of these sixty-four events can be assigned a numerical value in accordance with Definition 1.1.1. If the die is 'fair' then each of the elementary events $\{1\}, \{2\}, \ldots, \{6\}$, are equally likely so that by the argument used in the preceding example

$$P(\{1\}) = P(\{2\}) = \ldots = P(\{6\}) = \tfrac{1}{6}.$$

The probability of other events in $\mathscr{F}$ is obtained by decomposing them into a union of elementary events. For example if

$$\begin{aligned} A &= \text{'an even number is shown'} \\ &= \{2\} \cup \{4\} \cup \{6\} \end{aligned}$$

B

then

$$P(A) = P(\{2\})+P(\{4\})+P(\{6\}) = \tfrac{1}{2}. \qquad \square$$

In these two examples the actual values taken by the function P were determined by information extraneous to the space $(\Omega, \mathscr{F}, P)$, namely the supposition that the coin and die were fair. For the development of the theory such information is not necessary and in general we will not need to know any more about P than is implied by the conditions in its definition.

It is apparent that the decomposition of an event into a union of disjoint events is a useful device. Further results concerning relationships between events are contained in the Exercises at the end of this section. As a guide to intuition it is frequently helpful to use a

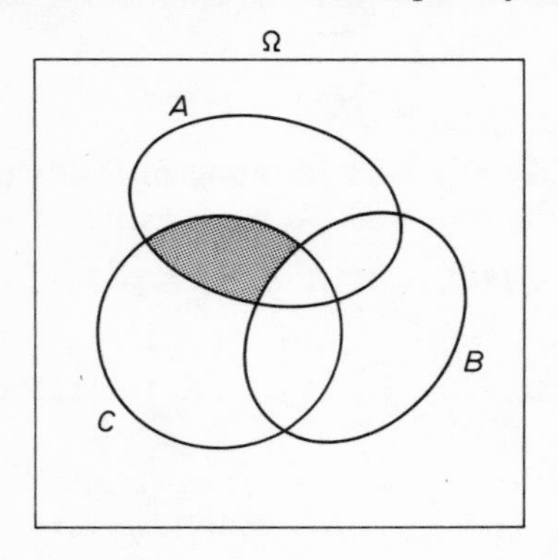

pictorial representation known as a *Venn diagram*. Suppose A, B, C are three events defined on Ω, and suppose that the elements of Ω are represented by the interior of the square in the figure. Being subsets of Ω each A, B, C, can be graphically represented as regions of this square. Thus the event $A\overline{B}C$ (i.e. the set of those ω which belong to both A and C but not to B) is represented by the shaded area.

As an example take the decomposition of $A \cup B$ into disjoint events. Considering only A and B in the figure it is clear that

$$A \cup B = A\overline{B} \cup \overline{A}B \cup AB$$

is such a decomposition. Also

$$A = AB \cup A\overline{B}, \qquad B = \overline{A}B \cup AB.$$

Hence

$$P(A \cup B) = P(A) + P(B) - P(AB)$$

as in (1.2.5).

Consider now two examples of probability spaces in which the sample space Ω contains infinitely many elements. We envisage the possibility of an observation on our underlying random phenomenon yielding one of an infinite number of alternatives.

Example 3 Suppose Ω contains a countably infinite number of elements $\Omega = \{\omega_0, \omega_1, \omega_2, \ldots\}$. Examples of such Ω are the set of natural numbers or the set of integers. As in the finite case let $\mathscr{F}$ contain $\varnothing$, Ω, and all the subsets (events) of Ω. It is apparent that $\mathscr{F}$ is closed under the formation of complements, countable unions and intersections and so is a σ field. There remains the probability function P. Recall that to each $A = \{\omega_{a_1}, \omega_{a_2}, \ldots\} \in \mathscr{F}$ the probability $P(A)$ is required to satisfy (i), (ii) and (iii)′ of Definition 1.1.1.

It is convenient to consider first the probability $P(\{\omega_j\})$ of elementary events. Let

$$P(\{\omega_j\}) = p_j, \qquad j = 0, 1, \ldots.$$

Then from the definition

$$0 \leqslant p_j \leqslant 1, \qquad \text{all } j,$$

and

$$1 = P(\Omega) = P\left(\bigcup_{j=0}^{\infty} \{\omega_j\}\right) = \sum_{j=0}^{\infty} P(\{\omega_j\}) = \sum_{j=0}^{\infty} p_j.$$

Hence, $p_0, p_1, p_2, \ldots$ are numbers lying in the unit interval $[0, 1]$ whose sum is unity. The probability of an event

$$A = \{\omega_{a_1}, \omega_{a_2}, \omega_{a_3}, \ldots\}$$

is given by the sum

$$P(A) = p_{a_1} + p_{a_2} + p_{a_3} + \ldots.$$

Since Ω is infinite, the σ field $\mathscr{F}$ contains infinitely many events. In particular suppose $A_1, A_2, A_3, \ldots$, are non-decreasing

$$A_1 \subseteq A_2 \subseteq A_3 \subseteq \ldots,$$

with limit defined by equation (A1.2) of the Appendix to this chapter as

$$\lim_{n\to\infty} A_n = \bigcup_{n=1}^{\infty} A_n.$$

By the theorem of the Appendix

$$\lim_{n\to\infty} P(A_n) = P(\lim_{n\to\infty} A_n) = P\left(\bigcup_{n=1}^{\infty} A_n\right),$$

the last probability being well defined since $\mathscr{F}$ is a σ field. A case of especial interest arises when

$$A_n = \{\omega_0, \omega_1, \ldots, \omega_n\}, \qquad n = 1, 2, \ldots.$$

For these A_n

$$P(A_n) = P\left(\bigcup_{j=0}^{n} \{\omega_j\}\right) = \sum_{j=0}^{n} p_j,$$

and, since $\lim_{n\to\infty} A_n = \Omega$, we recover

$$1 = P(\Omega) = \lim_{n\to\infty} P(A_n) = \lim_{n\to\infty} \sum_{j=0}^{n} p_j. \qquad \square$$

Sequences of numbers such as the above $\{p_j\}$ are of considerable importance and it is convenient to have a name for them.

Definition 1.2.1 If $\{p_x\}$ is a countable sequence of numbers such that $0 \leqslant p_x \leqslant 1$ for all x and $\sum_x p_x = 1$, then $\{p_x\}$ is called a *discrete probability distribution*. As a function of x the partial sum $\sum_{y \leqslant x} p_y$ is called a *discrete probability distribution function*. $\square$

We will frequently abbreviate terminology by referring merely to distributions or distribution functions. Note that if Ω is finite then only finitely many p_x are non-zero.

The procedure followed in Example 3 (which is readily applicable to the first two examples) was to obtain the probability distribution $\{p_j\}$ from the given probability space $(\Omega, \mathscr{F}, P)$ by $p_j = P(\{\omega_j\})$. The converse is also true in the sense that given a set of numbers $\{f_j\}$ which constitute a discrete probability distribution it is always possible to construct a triple $(\Omega', \mathscr{F}', P')$ satisfying the requirements that it be a probability space. To see this it is only necessary to take Ω' as the set of integers for which $\{f_j\}$ is non-zero,

$$\Omega' = \{j \mid f_j > 0\},$$

and to define the probability function P' by

$$P'(\{j\}) = f_j, \qquad j \in \Omega'.$$

It is easy to check that $(\Omega', \mathscr{F}, P')$ is a probability space and it is said to be *induced* by $\{f_j\}$. The point of these remarks is that whether our probability structure is specified via a space $(\Omega, \mathscr{F}, P)$ or a distribution $\{p_j\}$ is immaterial, as (certainly in the discrete case) we can pass from one form to the other without difficulty. These questions will be touched on again when we introduce random variables in § 1.4.

Example 4 Consider the technically more difficult situation when Ω is non-denumerable, say the real line $\Omega = \{\omega \mid -\infty < \omega < \infty\}$. Each real number is now an elementary event and the difficulty is due to the fact that the subsets of the real line are of a considerably more diverse nature than those of countable sets such as the integers.

To simplify matters consider the class of subsets of the real line *generated* by the semi-infinite intervals

$$A_x = \{(-\infty, x]\} = \{\omega \mid -\infty < \omega \leqslant x\}$$

as x runs through real values. We mean the following. First form the set $\mathscr{F}_1$ containing all complements, unions, and intersections of the A_x. Then form $\mathscr{F}_2$ containing all complements, unions, and intersections of members of $\mathscr{F}_1$. Continuing this procedure an arbitrarily large number of times we have in the limit the set $\mathscr{F}$ which is evidently closed under a countable number of set operations. This $\mathscr{F}$ is said to be the σ field *generated* by the A_x. That $\mathscr{F}$ contains Ω follows from the monotonicity of the A_x, since if $\{x_n\}$ is any sequence with $\lim_{n\to\infty} x_n = \infty$ then

$$\lim_{n\to\infty} A_{x_n} = \bigcup_{n=1}^{\infty} A_{x_n} = \{\omega \mid -\infty < x < \infty\} = \Omega. \qquad (1.2.6)$$

If P is a countably additive probability function satisfying Definition 1.1.1, the fact that all events in $\mathscr{F}$ can be obtained by the formation of complements, unions, and intersections of the A_x implies that it is sufficient to prescribe P only for the events A_x. That is, if the functional form of

$$P(A_x) = P(\{(-\infty, x]\}), \qquad -\infty < x < \infty,$$

is given then the probability of an arbitrary event in $\mathscr{F}$ can be written down in terms of this function. For example if $x_1 < x_2 < x_3 < x_4$ and

$$B = \{(-\infty, x_3]\} \cap \overline{\{(x_1, x_2]\} \cup \{(x_4, \infty)\}}$$

then

$$P(B) = P(A_{x_1}) + P(A_{x_3}) - P(A_{x_2}) + 1 - P(A_{x_4}).$$

Because of the monotonicity of the A_x, the theorem of the Appendix and (1.2.6) imply

$$\lim_{x\to\infty} P(A_x) = \lim_{n\to\infty} P(A_{x_n}) = P(\Omega) = 1.$$

Similarly if $y_1 \geqslant y_2 \geqslant y_3 \geqslant \ldots$ and $y_n \to -\infty$ then from (A1.1)

$$\begin{aligned} \lim_{x\to-\infty} P(A_x) &= \lim_{n\to\infty} P(A_{y_n}) \\ &= P\Big(\bigcap_{n=1}^{\infty} \{(-\infty, y_n]\}\Big) \\ &\leqslant P(\overline{\Omega}) = 0. \end{aligned}$$

From (1.2.4) it follows that $P(A_x)$ must be a monotonic function of x which varies from 0 to 1 as x increases from $-\infty$ to ∞.

To sum up, if Ω is the real line and $\mathscr{F}$ and P are chosen as above then the triple $(\Omega, \mathscr{F}, P)$ forms a probability space to which our theory applies. In fact the argument still goes through if Ω is a subset of the line (such as the integers) provided only that we can generate a σ field of events from a class of monotonic interval sets. □

It is important to note the fundamental role played by the interval sets $A_x = \{(-\infty, x]\}$ and their probabilities $P(A_x)$.

Definition 1.2.2 If $(\Omega, \mathscr{F}, P)$ is a probability space with Ω the real line (or a subset of the real line) and $\mathscr{F}$ the σ field generated by the sets $A_x = \{(-\infty, x]\}$, $-\infty < x < \infty$, the function

$$F(x) = P(A_x)$$

is called the *probability distribution function* associated with $(\Omega, \mathscr{F}, P)$.

□

Given the probability function P, the definition shows how to obtain the functional form of the corresponding distribution function $F(x)$. Conversely, if $F(x)$ is a given non-decreasing function of the real variable x with

$$\lim_{x \to -\infty} F(x) = 0, \qquad \lim_{x \to \infty} F(x) = 1,$$

we can induce a probability space $(\Omega', \mathscr{F}', P')$ with P' defined by

$$P'(\{(-\infty, x]\}) = F(x),$$

$\mathscr{F}'$ which is the σ field generated by the monotone sets

$$A_x = \{(-\infty, x]\}, \ -\infty < x < \infty,$$

and Ω' the real line. As in the countable case it is therefore immaterial whether our probability structure is prescribed by detailing $(\Omega, \mathscr{F}, P)$ or by giving the functional form of a distribution function $F(x)$. In the discrete case $F(x)$ is a step function, but in general a distribution function is required only to vary monotonically from 0 to 1 as x increases from $-\infty$ to ∞.

Finally suppose B_x is the interval set $B_x = \{(x, x+dx]\}$. For an increment dx the probability $P(B_x)$ is, in terms of the distribution function $F(x)$,

$$\begin{aligned} P(B_x) &= P(\{(-\infty, x+dx]\}) - P(\{(-\infty, x]\}) \\ &= F(x+dx) - F(x). \end{aligned}$$

In the case of countable $\Omega = \{\omega_0, \omega_1, \ldots\}$ this reduces to

$$P(B_x) = \sum_{x < j \leqslant x+dx} P(\{\omega_j\}).$$

which is zero if $j \notin (x, x+dx]$. In the non-denumerable case it may happen that the distribution function $F(x) = P(\{(-\infty, x]\})$ is a differentiable function of x with derivative $F'(x) = f(x)$, that is

$$\lim_{\delta \to 0} \frac{F(x+\delta) - F(x)}{\delta} = f(x).$$

If this is so we can write

$$F(x+\delta) = F(x) + \delta f(x) + o(\delta) \qquad \text{as } \delta \to 0,$$

and*

$$\begin{aligned} P(B_x) &= F(x+dx) - F(x) \\ &= f(x)\,dx + o(dx). \end{aligned}$$

Consequently if $F(x)$ is differentiable the probability $P(B_x)$ that our random phenomenon realises an observation in $(x, x+dx]$ is, to the first order of magnitude, proportional to the derivative of $F(x)$. Note that $F'(x) = f(x)$ is non-negative for all x (as $F(x)$ is non-decreasing) and since

$$F(x) = \int_{-\infty}^{x} f(y)\,dy,$$

a differentiable distribution function is uniquely determined by its derivative.

Definition 1.2.3. If $f(x)$, $-\infty < x < \infty$, is a non-negative function such that $\int_{-\infty}^{\infty} f(x)\,dx = 1$ then $f(x)$ is called a *probability density function.* The integral $F(x) = \int_{-\infty}^{x} f(y)\,dy$ is called a (*continuous*) *probability distribution function*. □

The basic notion is that of distribution function. The two important cases to which our discussion will be confined are

(i) when the distribution function $F(x)$ is a step function (sample space countable);
(ii) when $F(x)$ is differentiable with density $f(x) = F'(x)$.

The interpretation of the density is that when multiplied by dx it approximates to the first order the probability $P(\{(x, x+dx]\})$.

* $o(\delta)$ denotes a function of δ such that $\lim_{\delta \to 0} \delta^{-1}\, o(\delta) = 0$. The o and 0 notation is discussed in § 2.1.

Relevant issues will be taken up later and in the next section we return to the case when Ω contains only a finite number of points.

Exercises

In questions 1–6 A, B, C are events on a sample space Ω.

1. Verify the following:

(a) Commutative laws:

$$\begin{aligned} A \cup B &= B \cup A \\ AB &= BA. \end{aligned}$$

(b) Associative laws:

$$\begin{aligned} A \cup (B \cup C) &= (A \cup B) \cup C \\ A(BC) &= (AB)C. \end{aligned}$$

(c) Distributive laws:

$$\begin{aligned} A(B \cup C) &= (AB) \cup (AC) \\ A \cup (BC) &= (A \cup B)(A \cup C). \end{aligned}$$

2. Find expressions for

(a) Exactly one occurs.
(b) Exactly two occur.
(c) Only C occurs.
(d) At least one occurs.
(e) At most two occur.
(f) Both B and C, but not A, occur.
(g) None occur.
(h) Only A occurs.
(i) All three occur.

3. Establish the following:

(a) $\bar{A} \cup \bar{B} = \overline{(AB)}$.
(b) $\overline{(A \cup B)} = \bar{A}\bar{B}$.
[Note that in general $\overline{A \cup B} \neq \bar{A} \cup \bar{B}$ and $\bar{A}\bar{B} \neq \overline{AB}$].
(c) $(AB) \cup C = C \cup (AB\bar{C})$ [thus $(AB) \cup C \neq AC \cup BC$, c.f. 1(c)].
(d) $(A \cup B) - B = A\bar{B}$.

(e) $(\overline{\bar{A}\cup\bar{B}}) = AB$.

(f) The general form of (a) and (b) is frequently called *de Morgan's law* and can be expressed as $\bigcup_{j=1}^{n} A_j = \overline{\left(\bigcap_{j=1}^{n} \bar{A}_j\right)}$.

4. By writing $A\cup B\cup C$ as a union of disjoint sets obtain the generalisation of (1.2.5):

$$P(A\cup B\cup C) = P(A) + P(B) + P(C) - P(AB) - P(AC) - P(BC) + P(ABC).$$

5. Interpret $A\bar{B}\cup\bar{A}B$ as the event that exactly one of A or B occurs and show that

$$P(A\bar{B}\cup\bar{A}B) = P(A)+P(B)-2P(AB).$$

[Contrast this result with (1.2.5)].

6. Establish the inequalities

$$P(ABC) \leqslant P(AB) \leqslant P(A\cup B) \leqslant P(A\cup B\cup C) \leqslant P(A)+P(B)+P(C).$$

7. Given the events $A_1, A_2, \ldots, A_n$, use an inductive argument to establish *Boole's inequality*

$$P\left(\bigcup_{j=1}^{n} A_j\right) \leqslant \sum_{j=1}^{n} P(A_j).$$

8. The set $\mathscr{F}$ was defined to be a field if it was closed under the formation of unions and complements (page 3). Show that this implies that $\mathscr{F}$ is also closed under the formation of intersections. [*Hint*: Use 3(e) above].

9. Let Ω be the real line and $\{x_n\}$ a doubly monotone sequence of numbers

$$\ldots \leqslant x_{-1} \leqslant x_0 \leqslant x_1 \leqslant \ldots.$$

If $B_j = \overline{\{-\infty, x_{-j}]\}}\cap\{(-\infty, x_j]\}$ write down the probabilities of the following events in terms of the distribution function

$$F(x) = P(\{(-\infty, x]\}):$$

(a) $\overline{B}_1 B_3 \cup \overline{B}_3 B_5$, (b) $\bigcup_{j=1}^{n} \overline{B}_j B_{j+1}$, (c) $(B_1 \cup \overline{B}_2) B_3$,

(d) $\lim_{n\to\infty} \bigcup_{j=1}^{n} B_j$ (e) $\lim_{n\to\infty} \bigcap_{j=1}^{n} B_j$.

10. Two dice are rolled. Specify the events

(a) the sum of the faces is $\leqslant 5$,
(b) the sum of the two faces is an even number,
(c) the difference of the two faces is equal to three.

11. Two numbers are chosen from the set $\{1, 2, 3, 4\}$. Specify

(a) the sample space,
(b) the events 'the sum of the chosen numbers is 5' and 'the first number is greater than the second'.

12.(a) If a sample space can be written as $\Omega = A \cup B$ where $P(A) = \frac{1}{2}$, $P(B) = \frac{3}{4}$, find $P(AB)$.

(b) If $\Omega = \{1, \frac{1}{2}, (\frac{1}{2})^2, \ldots, (\frac{1}{2})^n\}$, n fixed, and

$$A = \{1, \tfrac{1}{2}\}, \qquad B = \{\omega \mid \omega = (\tfrac{1}{2})^j, j \geqslant 1 \text{ and even}\}$$

with $P(A) = \frac{1}{8}$, $P(B) = \frac{1}{2}$, find $P(A \cup B)$ and $P(\overline{A}\overline{B})$.

13. Verify that each of the following $\{p_j\}$ is a discrete probability distribution and aescribe the induced probability space.

(a) $p_j = e^{-\lambda} \dfrac{\lambda^j}{j!}$, $\quad j = 0, 1, 2, \ldots, \lambda > 0$.

(b) $p_j = \dbinom{N}{j} p^j (1-p)^{N-j}$, $\quad j = 0, 1, 2, \ldots, N; \quad 0 < p < 1$.

(c) $p_j = m^{-1}$, $\quad j = 1, 2, \ldots, m$.

14. Check that the following are probability density functions.

(a) $f(x) = 1, \quad 0 < x < 1,$
$\qquad = 0 \quad \text{otherwise.}$

(b) $f(x) = \dfrac{1}{\pi(x(1-x))^{\frac{1}{2}}}, \quad 0 < x < 1,$
$\qquad = 0 \quad \text{otherwise.}$

(c) $f(x) = \lambda e^{-\lambda x}, \qquad 0 < x, \lambda > 0,$
$\qquad = 0 \quad \text{otherwise.}$

15. Verify that $F(x)$ defined below is a probability distribution function for $\lambda > 0$.

$$\begin{aligned} F(x) &= 0, & x &\leqslant -1, \\ &= \tfrac{1}{4}, & -1 < x &\leqslant -\tfrac{1}{2}, \\ &= \tfrac{1}{2}, & -\tfrac{1}{2} < x &\leqslant 0, \\ &= 1-\tfrac{1}{2}e^{-\lambda x}, & 0 &< x. \end{aligned}$$

Construct a probability space induced by this distribution function.

16. An urn contains N balls of which M are red. A ball is drawn at random and its colour noted before it is replaced. The procedure is continued until a red ball is obtained for the first time. If the chance phenomenon of interest is the number of such drawings, what is the sample space and what is the probability that m are required? Are the assertions

(a) at least two drawings required,
(b) between three and six drawings required,
events on the space? If so, find their probabilities.

1.3 FINITE PROBABILITY SPACES

Suppose Ω contains only m elements $\Omega = \{\omega_1, \omega_2, \ldots, \omega_m\}$.

Our first result is that the field of events $\mathscr{F}$ defined on Ω contains exactly 2^m members. This can be shown by a combinatorial argument which depends on the fact that the number of different sets of r

objects which can be chosen from a collection of m distinct objects is

$$\frac{m(m-1)\ldots(m-r+1)}{r!} = \binom{m}{r}. \tag{1.3.1}$$

Indeed there are m ways of choosing the first object; having made this choice there are $m-1$ ways of choosing the second; ...; and finally $m-r+1$ ways of choosing the rth. Then the total number of ways taking order into account is

$$m(m-1)\ldots(m-r+1) = (m)_r. \tag{1.3.2}$$

If order is not important (i.e. if one is interested in the number of sets as opposed to the number of vectors with distinct components) it is necessary to divide this product by $r(r-1)\ldots 2.1 = r!$, since to each set of r objects $\{a_1, a_2, \ldots, a_r\}$ there correspond $r!$ vectors, namely

$$(a_1, a_2, \ldots, a_r), (a_2, a_1, \ldots, a_r), \ldots, (a_r, a_{r-1}, \ldots, a_1).$$

Hence the number of subsets of Ω containing exactly r elements is

$$(m)_r(r!)^{-1} = \binom{m}{r}.$$

But $\mathscr{F}$ contains these subsets for $r = 1, 2, \ldots, m-1$, in addition to $\varnothing$ (corresponding to $r = 0$) and Ω (corresponding to $r = m$). Therefore the total number of events is

$$\sum_{r=0}^{m} \binom{m}{r} = (1+1)^m = 2^m$$

by the binomial theorem.

The probability function P assigns a numerical value to each of these 2^m events in the manner prescribed by Definition 1.1.1. However, as pointed out when discussing examples of discrete probability spaces in the preceding section it is usually more convenient to define P by allotting numerical values to only the m probabilities $P(\{\omega_j\})$, $j = 1, 2, \ldots, m$, of elementary events rather than to the 2^m probabilities $P(A_j), j = 1, 2, \ldots, 2^m$, of events. Given

$$P(\{\omega_j\}) = p_j, \qquad j = 1, 2, \ldots, m,$$

and if $A_j = \{\omega_{j_1}, \omega_{j_2}, \ldots, \omega_{j_r}\}$ is any event in $\mathscr{F}$, then

$$\begin{aligned} P(A_j) &= P(\{\omega_{j_1}\}) + (\{\omega_{j_2}\}) + \ldots + P(\{\omega_{j_r}\}) \\ &= p_{j_1} + p_{j_2} + \ldots + p_{j_r}. \end{aligned}$$

The important special case of 'equally likely outcomes' refers to *equally likely elementary events* (as in tossing a fair coin or rolling a fair die). In this case

$$P(\{\omega_j\}) = p_j = m^{-1}, \qquad j = 1, 2, \ldots, m,$$

and if $A_j = \{\omega_{j_1}, \omega_{j_2}, \ldots, \omega_{j_r}\}$ then

$$P(A_j) = rm^{-1} = \frac{\text{number of elementary events in } A_j}{\text{number of elementary events in } \Omega}. \qquad (1.3.3)$$

Hence if the elementary events are equally likely the probability of any event $A \in \mathscr{F}$ is obtained from (1.3.3) by counting the number of elementary events in A. Making this count often involves some combinatorial analysis and, in particular, the frequent use of (1.3.1) and (1.3.2).

Example 1 A card is drawn from a shuffled pack of fifty-two. What is the probability that it is an ace? The sample space consists of fifty-two elements, one for each card. Since the pack is well shuffled by definition we can assume that these elementary events are equally likely. The number of elementary events favourable to drawing an ace (i.e. the number of ω_i in the event 'drawn card is an ace') is four. By (1.3.3) the required probability is $4/52 = 1/13$. □

Example 2 Two cards are drawn from a shuffled pack. What is the probability that both are aces? An elementary event consists of a pair $\{a, b\}$ and the number of such pairs that can be formed from fifty-two distinct objects is $\binom{52}{2}$ (by 1.3.1). By definition these $\binom{52}{2}$ elementary events are equally likely. The number of elementary events favourable to drawing two aces is equal the number of ways of choosing a set of two objects out of four (since there are four aces), namely $\binom{4}{2}$. The answer is $\binom{4}{2}\Big/\binom{52}{2} = (221)^{-1}$. □

Note that the sample space depends on the question asked and to an extent on the method used to answer it. If the fifty-two cards are numbered serially $1, 2, \ldots, 52$ then in the first example

$$\Omega = \{j \mid j = 1, 2, \ldots, 52\}$$

and in the second $\Omega = \{\{a, b\} \mid a, b = 1, 2, \ldots, 52, a \neq b\}$. Furthermore in Example 2 an alternative sample space is the set

$$\{(a, b) \mid a, b = 1, 2, \ldots, 52, a \neq \mathrm{b}\}$$

of $(52)_2$ vectors, the favourable cases now numbering $(4)_2$.

The argument used to obtain (1.3.1) as the number of ways of choosing a set of r objects from m depended on the fact that each object could be chosen only once. This method of choice is sometimes called *sampling without replacement.* Since we have been writing m for the number of sample points, let us alter the notation and suppose that we have N distinguishable objects. Then by the argument for (1.3.1) the number of distinct sets of size n that can be obtained when sampling without replacement is $\binom{N}{n}$. Put another way, if repetitions are not permitted, $\binom{N}{n}$ is the number of different samples of n from N.

A case of some importance arises from repeated application of this reasoning. Suppose the N objects can be subdivided into two groups, say 'red' and 'white', consisting of M and $N-M$ individuals respectively. A random sample of n is chosen without replacement, and we ask for the probability that these n consist of m red and $n - m$ white objects.

The sample of n can be chosen in $\binom{N}{n}$ ways, and this is the number of elementary events. To compute the number of favourable outcomes we require the number of sets of size n which contain m red and $n - m$ white. But this is

$$\binom{M}{m}\binom{N-M}{n-m} \tag{1.3.4}$$

since the m red can be chosen in $\binom{M}{m}$ ways, each of these being capable of selection with each of the $\binom{N-M}{n-m}$ possible sets of $n-m$ white objects. The required probability is

$$h_m = \binom{M}{m}\binom{N-M}{n-m}\Big/\binom{N}{n}. \tag{1.3.5}$$

Exercise 7 at the end of this section requests that

$$\{h_m, m = 0, 1, 2, \ldots, \min(n, M)\}$$

be shown to be a probability distribution. Because of its frequent occurrence it is given a special name and is known as the *hyper-geometric distribution.*

It is instructive to consider the same problem modified by changing the sampling scheme to be with replacement. That is, we have M red and $N-M$ white objects of which n are randomly chosen by sampling with replacement. Repetitions are allowed and we ask for the probability that the sample of n consists of m red and $n-m$ white under this procedure.

Since we are sampling with replacement there are N ways of choosing the first object and indeed N ways of making any of n choices. The total number of vectors (i.e. ordered sets) is therefore N^n. Of these vectors the number that have red objects in *m specified* coordinates and white in the other $n - m$ places is $M^m(N - M)^{n-m}$ However to each set of m red and $n-m$ white there correspond $\binom{n}{m}$ such vectors, since this is the number of ways in which m places can be chosen from n. The number of favourable vectors is therefore $\binom{n}{m}M^m(N-M)^{n-m}$ and the required probability is

$$\begin{aligned} b_m &= \binom{n}{m}\frac{M^m(N-M)^{n-m}}{N^n} \\ &= \binom{n}{m}\left[\frac{M}{N}\right]^m\left[1-\frac{M}{N}\right]^{m-n}. \end{aligned} \tag{1.3.6}$$

By the binomial theorem

$$\sum_{m=0}^{n} b_m = \left[\frac{M}{N}+1-\frac{M}{N}\right]^n = 1$$

and hence $\{b_m, m = 0, 1, \ldots, n\}$ is a probability distribution, called the *binomial distribution.* It plays an important role in the theory and reappears frequently in subsequent chapters. For the moment the important point is that the two sampling procedures, without or with replacement, lead to the two different probability distributions (1.3.5) and (1.3.6). (see Exercise 9).

Example 3 A fair die is rolled three times in succession. What is the probability that six appeared on the first two throws and five on the third? We regard the number on each face of the die as a distinguishable object so that $N = 6$. Rolling the die three times corresponds to choosing $n = 3$ of these six objects when sampling with replacement. Our question involves order so that we want the number of three component vectors that can be formed from six objects, allowing repetitions. This is 6^3 and the sample space contains the 6^3 elements $\Omega = \{(a, b, c) \mid a, b, c = 1, 2, \ldots, 6\}$. The event 'six on first two throws, five on the third' contains only one elementary event, namely $(6, 6, 5)$. Since these events are equally likely we have from (1.3.3) that the required probability is $1/6^3$.

Exercise Continue Example 3 by finding the probability of the following events:

(a) two sixes and one five are shown;
(b) the same number appears on each throw;
(c) at least two sixes are thrown;
(d) the sum of the three numbers thrown is equal to seven. □

Example 4 Random digits. Consider the $N = 10$ digits $0, 1, 2, \ldots, 9$, and suppose four digits are consecutively chosen at random. If repetitions are allowed, $N^n = 10^4$ different numbers can be formed. On the other hand only $(N)_n = (10)_4$ different numbers can be obtained if we require each digit to be different. It follows that the probability of all four successive random digits being different is

$$\frac{(10)_4}{10^4} = \frac{10.9.8.7}{10^4} = 0{\cdot}504. \qquad \square$$

Example 5 A typical situation in which the hypergeometric distribution is applicable concerns the game of bridge. The number of hands that can be dealt to a player is $\binom{52}{13}$, since a hand is a sample of thirteen cards from fifty two drawn without replacement. The probability that a hand dealt from a well shuffled pack contains, say, seven clubs is, from (1.3.5),

$$\binom{13}{7}\binom{39}{6}\Big/\binom{52}{13}.$$

By way of contrast the probability of obtaining seven clubs if the sample of thirteen had been drawn with replacement is the binomial probability

$$\binom{13}{7}\left[\frac{13}{52}\right]^7\left[\frac{39}{52}\right]^6.$$

Note that 13/52 is the proportion of clubs in the pack and is the probability of obtaining a club on any particular draw. □

We have confined our attention in this section so far to sample spaces with equally likely outcomes and have seen that questions of a counting or combinatorial nature are all important. Combinatorial analysis has played an important historical role in the development of probability theory, as in the past sample spaces with this property were central to the subject. The more modern axiomatic treatment* does not pay particular attention to these sample spaces although combinatorial arguments still play a very important role in certain branches. For further discussion of combinatorial problems in probability theory we refer to the relevant chapters of the books by Feller and Parzen cited in the preface.

Turning to finite sample spaces without necessarily equally likely elementary events we mention only that, whilst these probability spaces do have certain special properties, it is more convenient to treat them as particularisations of the countable case. For this reason we do not develop a separate theory.

Exercises

1. Two distinguishable dice are thrown. Write out at length all

* Dating from the first decade of this century and formalised by the publication (in German) of ***Foundations of the Theory of Probability*** by the Russian mathematician A. N. KOLMOGOROV in 1933.

thirty-six members of the sample space. Which of these are elements of the events

(a) the first die shows two,
(b) the second die shows two,
(c) the two dice show equal numbers,
(d) the sum of the two numbers is $\geqslant 12$?

2. (Continuation.) Specify the elementary events contained in

(a) two appears,
(b) the sum of the two numbers is even,
(c) the sum of the two numbers is $\leqslant 5$,
(d) the two numbers differ by exactly three.

3. (Continuation) Assuming the dice are fair, give the probabilities of the events appearing in Exercises 1 and 2.

4. State how many n-letter words can be formed from an alphabet of N symbols. Referring to the Latin alphabet how many five-letter words can be constructed in which the second letter is a vowel?

5. Two balls are drawn without replacement from an urn containing three red and two white balls. Compute the probabilities that:

(a) both balls are of the same colour,
(b) both balls are red,
(c) the first ball drawn is red, the second white,
(d) at least one red ball is drawn.

6. Rework Exercise 5 under the assumption that the sampling is done with replacement.

7. Show that

$$\sum_{m=0}^{\min(n,M)} \binom{M}{m}\binom{N-M}{n-m} = \binom{N}{n}.$$

Hence (1.3.5) does in fact define a probability distribution.
Hint: Use induction or else power series thus:

$$(1+z)^N = (1+z)^M(1+z)^{N-M}.$$

Expand both sides and equate coefficients. Related material is in § 2.5. (ii)].

8. An urn contains ten balls of which five are red. What is the probability that all of a sample of three balls are red when the sample is

(a) drawn with replacement,
(b) drawn without replacement

9. Show that for large N the hypergeometric probabilities (1.3.5) approach those of the binomial, (1.3.6), i.e. show

$$\lim_{N\to\infty} h_m = b_m, \qquad m = 0, 1, 2, \ldots, n.$$

The implication is that when the population from which objects are being selected is large, the distinction between sampling with or without replacement becomes unimportant. In this case one uses the simpler binomial formula (1.3.6).

10. Extend (1.3.5) as follows. Suppose the N objects are subdivided into k groups consisting of

$$N_1, N_2, \ldots, N_{k-1}, N-(N_1+N_2+\ldots+N_{k-1}),$$

individuals respectively. Show that if a random sample of n is taken without replacement then the probability of obtaining n_j from the jth group, $j = 1, 2, \ldots, k$, is

$$\binom{N_1}{n_1}\binom{N_2}{n_2}\ldots\binom{N_{k-1}}{n_{k-1}}\binom{N-N_1-N_2\ldots-N_{k-1}}{n-n_1-n_2\ldots-n_{k-1}}\Big/\binom{N}{n}$$

11. Extend the binomial formula (1.3.6) as follows. Suppose of N objects, N_1 are red, N_2 white, and $N-(N_1+N_2)$ purple. A sample of n is taken with replacement and it is required to find the probability b_{n_1,n_2} that the sample contains n_1 red, n_2 white, and $n-(n_1+n_2)$ purple objects. Using the fact that

$$\binom{n}{n_1}\binom{n-n_1}{n_2} = \frac{n!}{n_1!n_2!(n-n_1-n_2)!}$$

show that

$$b_{n_1,n_2} = \frac{n!}{n_1!n_2!(n - n_1 - n_2)!} \times \left[\frac{N_1}{N}\right]^{n_1} \left[\frac{N_2}{N}\right]^{n_2} \left[1 - \left\{\frac{N_1 + N_2}{N}\right\}\right]^{n-(n_1+n_2)}$$

Generalise this to the case when N can be divided into k subgroups. The expression b_{n_1,n_2} and its generalisation define the *multinomial distribution.*

The following two exercises refer to bridge.

12. Show that

(a) the number of hands containing two aces is $\binom{4}{2}\binom{48}{11}$;

(b) the number of different ways in which a pack can be dealt into four hands is $\binom{52}{13}\binom{39}{13}\binom{26}{13}$;

(c) the number of hands containing the ace of spades is $\binom{51}{12}$;

(d) the number of hands containing ten specified cards is $\binom{42}{3}$.

13. Find the probability that

(a) North has all four aces,
(b) a hand contains four clubs, one diamond, six hearts, and two spades,
(c) a hand contains four specified clubs, one specified diamond, six specified hearts, and two specified spades.
(d) North and South between them have all four aces.

14. The calculation of binomial coefficients becomes tedious when the numbers concerned are large. *Stirling's formula* states that

$$\lim_{n\to\infty} \frac{n!}{(2\pi)^{\frac{1}{2}} n^{n+\frac{1}{2}} e^{-n}} = 1.$$

For a proof see for example FELLER I page 52. For large n an approximation to $n!$ is therefore given by the denominator on the left-hand side. Use this result to approximate the quantities in Exercise 12.

1.4 RANDOM VARIABLES

Suppose a probability space $(\Omega, \mathscr{F}, P)$ is given. We commence with the relatively simple case in which Ω is discrete, containing at most a countably infinite number of elementary events.

Definition 1.4.1 A real valued function defined on a discrete sample space is called a *(discrete) random variable.* □

A random variable X is then a function such that for each $\omega \in \Omega$, $X(\omega)$ is a real number. Since Ω is discrete, X can realise at most a countable number of distinct values and considered as a mapping X is from Ω into a countable subset of the real line. Discreteness also implies that $\mathscr{F}$ contains all subsets of Ω, in particular those of the form $\{\omega \mid X(\omega) \leqslant x\}$.

Example 1 Toss of a coin, $\Omega = \{H, T\}$. Let X be the random variable defined by

$$X(H) = 1, \quad \text{i.e. } X = 1 \text{ if 'Heads' occur,}$$
$$X(T) = 0, \quad \text{i.e. } X = 0 \text{ if 'Tails' occur.}$$

If the coin is 'fair' the probability function P is given by

$$P(\{H\}) = P(\{T\}) = \tfrac{1}{2},$$

and we can speak of the *probability distribution of the random variable* X as

$$\Pr(X = 1) = P(\{H\}) = \tfrac{1}{2},$$
$$\Pr(X = 0) = P(\{T\}) = \tfrac{1}{2}.$$

It is of course possible to introduce numerous other random variables on this space, e.g.

$$Y(H) = 10^6, \qquad Y(T) = -\pi.$$ □

Example 2 Roll of a die, $\Omega = \{1, 2, 3, 4, 5, 6\}$. Let

$$X(\omega) = \omega, \qquad \omega = 1, 2, 3, 4, 5, 6,$$

i.e. the random variable X in this case denotes the face shown by rolling the die. Its probability distribution is

$$\Pr(X = x) = P(\{x\}), \qquad x = 1, 2, \ldots, 6.$$ □

Example 3 $\Omega = \{\omega \mid \omega = 0, \pm 1, \pm 2, \ldots\}$. Let

$$\begin{aligned} X(\omega) &= a, \qquad \omega \leqslant 0, \\ &= b, \qquad \omega > 0. \end{aligned}$$

In words, X takes the value b if the random phenomenon realises a positive value and is equal to a otherwise. Its probability distribution is

$$\Pr(X = a) = \sum_{j=-\infty}^{0} P(\{j\}),$$

$$\Pr(X = b) = \sum_{j=1}^{\infty} P(\{j\}).$$

Another random variable defined on this space is for example

$$Y(\omega) = \omega^2, \qquad \omega \in \Omega$$

with distribution

$$\Pr(Y = 0) = P(\{0\}), \quad \Pr(Y = j^2) = P(\{j\}) + P(\{-j\}), \quad j = 1, 2, \ldots$$

and $\Pr(Y = k) = 0$ otherwise. □

Example 4 Two distinguishable dice are rolled,

$$\Omega = \{(i,j) \mid i,j = 1, 2, \ldots, 6\}.$$

Let X denote the sum of the two numbers obtained.

$$X((i,j)) = i + j, \qquad i,j = 1, 2, \ldots, 6.$$

This example is continued in Exercises 2 and 3. □

In these examples the statement 'the probability that the random variable X is equal to x', written $\Pr(X = x)$, is given a numerical value by the probability function P, namely $\Pr(X = x)$ is the P value assigned to the event $\{\omega \mid X(\omega) = x\}$. If x_0 is a number such that $X(\omega) = x_0$ for no $\omega \in \Omega$ then $\{\omega \mid X(\omega) = x_0\} \subseteq \varnothing$ and from (1.2.4)

$$\Pr(X = x_0) \leqslant P(\varnothing) = 0.$$

The discrete nature of Ω thus entails that the set

$$\{\Pr(X = x), -\infty < x < \infty\}$$

in fact contains at most a countably infinite number of non-zero terms.

Definition 1.4.2 If X is a random variable on a discrete probability space $(\Omega, \mathscr{F}, P)$ the probability $\Pr(X = x)$ is defined by

$$\Pr(X = x) = P(\{\omega \,|\, X(\omega) = x) = p_x.$$

The sequence $\{p_x, -\infty < x < \infty\}$ of non-zero terms is called the *probability distribution of* X. As a function of x the partial sum $\sum_{y \leqslant x} p_y = \Pr(X \leqslant x)$ is called the *probability distribution function of* X. □

A reference back to Definition 1.2.1 indicates that the terms 'probability distribution' and 'probability distribution function' will be used inconsistently unless $0 \leqslant \Pr(X = x) \leqslant 1$ (which is obvious) and $\sum_x \Pr(X = x) = 1$. The latter requirement is equivalent to

$$\lim_{x \to \infty} \Pr(X \leqslant x) = 1$$

and that this is indeed the case is shown by the following argument:

$$\begin{aligned}\Pr(X \leqslant x) &= \sum_{y \leqslant x} P(\{\omega \,|\, X(\omega) = y\}) \\ &= P(\bigcup_{y \leqslant x} \{\omega \,|\, X(\omega) = y\}) \\ &= P(\{\omega \,|\, -\infty < X(\omega) \leqslant x\}).\end{aligned}$$

But the sets $A_x = \{\omega \,|\, -\infty < X(\omega) \leqslant x\}$ are non-decreasing in x. Hence by the theorem of the Appendix

$$\lim_{x \to \infty} \Pr(X \leqslant x) = \lim_{x \to \infty} P(A_x) = P(\lim_{x \to \infty} A_x)$$

$$= P(\{\omega| -\infty < X(\omega) < \infty\}) = P(\Omega) = 1.$$

We therefore have

Theorem 1.4.1 The quantities $\Pr(X = x) = P(\{\omega \,|\, X(\omega) = x\})$ of Definition 1.4.2 constitute a probability distribution in the sense of Definition 1.2.1, namely $0 \leqslant \Pr(X = x) \leqslant 1$, all x, and

$$\sum_x \Pr(X = x) = 1.$$ □

Random variables are usually specified by assigning numerical values to the probabilities $\Pr(X = x)$. For example the statement that the random variable X is distributed according to $\{p_x, x \in \mathscr{X}\}$, where $\mathscr{X}$ is some given countable set of real numbers, means simply that $\{p_x\}$ is a probability distribution and

$$\Pr(X = x) = p_x, \qquad x \in \mathscr{X},$$
$$\Pr(X = x) = 0, \qquad x \notin \mathscr{X}.$$

We have seen in § 1.2 that a probability distribution can be defined for any given (countable) probability space $(\Omega, \mathscr{F}, P)$ and that, conversely, given a probability distribution $\{p_x\}$ one can induce a probability space $(\Omega', \mathscr{F}', P')$. This implies that either approach can be used to specify the probability structure pertinent to a given random phenomenon and a valid question is then why introduce the seemingly redundant notion of a random variable? The answer is essentially that it is convenient to do so and the concept is essentially an aid to the cogent formulation of probabilistic problems.

A somewhat frivolous illustration is the proposition: 'an undergraduate met by chance has twenty cents in his pocket'. In the absence of further information one can regard the amount of money carried by an undergraduate as a random phenomenon with possible values (sample space) in cents $\{0, 1, 2, \ldots\}$. As a matter of terminology it is convenient to be able to assert that the money (in cents) carried by an undergraduate is a random variable X distributed over the non-negative integers. Here $X(\omega) = \omega$ and realisations of our random phenomenon are identical with realisations of our random variable. The functional form of the probabilities

$$\Pr(X = x), \qquad x = 0, 1, 2, \ldots,$$

is unknown but one nevertheless always has $0 \leqslant \Pr(X = x) \leqslant 1$ and $\sum_{x=0}^{\infty} \Pr(X = x) = 1$, thus defining a probability distribution and consequently a probability space.

So far we have assumed that Ω is countable. In this case $\mathscr{F}$ contains all subsets of Ω and Definition 1.4.1 imposed no restriction on the sorts of function X that could be random variables, since all the sets of the form $\{\omega \mid X(\omega) = x\} \in \mathscr{F}$ and hence could be assigned a P value. In the non-denumerable case we have to insist that the only subsets of Ω to be considered by the theory are those which are included in a σ field on which a probability function P is defined. If, as before, we are to define a random variable X as a function from Ω into the real line, we must restrict our class of functions to ensure that sets of the form $\{\omega \mid X(\omega) \leqslant x\}$ are members of $\mathscr{F}$;

for otherwise it would be impossible to attach a meaning to probabilistic statements about X.

Example 5 (c.f. Example 3) $\Omega = \{\omega \mid -\infty < \omega < \infty\}$. Let

$$\begin{aligned} X(\omega) &= a, \qquad \omega \leqslant 0, \\ &= b, \qquad \omega > 0. \end{aligned}$$

The inverse image X^{-1} is

$$\begin{aligned} X^{-1} &= \{\omega \mid \omega \leqslant 0\} = \{(-\infty, 0]\} \qquad \text{if } X = a, \\ &= \{\omega \mid \omega > 0\} = \{(0, \infty)\} \qquad \text{if } X = b. \end{aligned}$$

Both $\{(-\infty, 0]\}$ $\{(0, \infty)\}$ are members of $\mathscr{F}$ and the probability distribution of X is well defined by

$$\begin{aligned} \Pr(X = a) &= P(\{(-\infty, 0]\}), \\ \Pr(X = b) &= P(\{[0, \infty)\}). \end{aligned}$$ □

In general we have

Definition 1.4.3 Suppose $(\Omega, \mathscr{F}, P)$ is a probability space in which Ω is not necessarily countable. A *random variable* X defined on this space is a function from Ω into the real line such that the set $\{\omega \mid X(\omega) \leqslant x\} \in \mathscr{F}$ for every real x. The *probability distribution function of* X is defined to be

$$\Pr(X \leqslant x) = P(\{(-\infty, x]\}).$$

If $\Pr(X \leqslant x) = F(x)$ is differentiable, the derivative $F'(x) = f(x)$ is called the *probability density function of* X. □

An important special case is when $X(\omega) = \omega$ and Ω is the real line. If this is so,

$$F(x) = \Pr(X \leqslant x) = P(\{\omega \mid \omega \leqslant x\})$$

and, as in the countable case, the specification of a distribution function is sufficient to construct an induced probability space.

For X to be a random variable the sole requirement of Definition 1.4.3 is that probabilities can be assigned to the sets $\{\omega \mid X(\omega) \leqslant x\}$. Subject only to varying monotonically between 0 and 1, the distribution function $F(x) = P(\{\omega \mid X(\omega) \leqslant x\})$ may be a step function, or a differentiable function, or continuous but not differentiable, or indeed a mixture of all three. The general theory must be adequate to these different possibilities, but it is sufficient for our purposes to confine attention to the first two. *We will therefore assume that all random variables discussed are one or other of the following two types*:

(i) *Lattice variables*, which are those that can take only values of the form $b+kh$, $k = 0, \pm 1, \pm 2, \ldots$, with positive probability. The distribution function is then a step function which increases only at these points. If h is the largest positive number for which this is true it is called the *span* of the distribution function or the *span* of the random variable.

(ii) *Non-lattice variables with differentiable distribution functions.* These are random variables which vary continuously and possess density functions.

The lattice variables usually considered (e.g. Examples 1, 2, 4) take $b = 0$ and span $h = 1$. More generally if X is lattice with span h then $\Pr(X = x)$ is zero (or undefined) unless

$$x = b+kh, \qquad k = 0, \pm 1, \pm 2, \ldots.$$

We will on occasion use the terms lattice and discrete interchangeably, although this is not strictly correct.

The second category of random variables (which by an abuse of terminology are sometimes referred to as *absolutely continuous*) have distribution functions expressible as

$$\Pr(X \leqslant x) = F(x) = \int_{-\infty}^{x} f(y)\, dy.$$

The probability that such an X realises a value in the interval $(x, x+dx]$ is

$$\begin{aligned} \Pr(x < X \leqslant x+dx) &= F(x+dx)-F(x) \\ &= f(x)\, dx+o(dx), \qquad \text{as } dx \to 0. \end{aligned}$$

The value of the density function at x is then to the first order of magnitude proportional to the probability of the corresponding absolutely continuous random variable realising a value in an arbitrarily small interval about the point x (see also the discussion towards the end of § 1.2).

Example 6 Let $\Omega = \{\omega \mid -\infty < \omega < \infty\}$, $\mathscr{F}$ be the σ field generated by $\{(-\infty, x] \mid -\infty < x < \infty\}$, and P be defined by

$$P(\{(-\infty, x]\}) = F(x)$$

where

$$\frac{dF(x)}{dx} = f(x) = \frac{1}{(2\pi)^{\frac{1}{2}}} e^{-\frac{1}{2}x^2}, \qquad -\infty < x < \infty.$$

This distribution function $F(x)$ is known as the *normal* or *Gaussian* distribution function. Random variables with this distribution are said to be *normally distributed.* These will reappear frequently in the sequel. That $\int_{-\infty}^{\infty} f(x)\,dx = 1$ is shown in § 2.5. □

The general form of the so-called normal density is

$$f(x) = (\sigma^2 2\pi)^{-\frac{1}{2}} \exp\{-(x-\mu)^2/(2\sigma^2)\}, \qquad -\infty < x < \infty,$$

indexed by the two parameters μ and σ^2, $-\infty < \mu < \infty$, $0 < \sigma^2$. If X has this density we will often write

$$X \frown N(\mu, \sigma^2)$$

which translates as 'X is normally distributed with parameters μ and σ^2'.

Example 7 Let Ω and $\mathscr{F}$ be as in the preceding example and define P by

$$\begin{aligned} P(\{(-\infty, x]\}) &= 0, & x &\leqslant 0, \\ &= 1 - e^{-\lambda x}, & x &> 0, \end{aligned}$$

with λ some positive constant. If the random variable X is defined by the distribution function $F(x) = \Pr(X \leqslant x) = P(\{(-\infty, x]\})$, then because of the functional form of $F(x)$, X is said to obey the *negative exponential distribution.* □

Exercises

1. The following are verbal descriptions of random variables:

   (a) the number of aces in a bridge hand,
   (b) the error made when measuring a given physical constant,
   (c) the number of tosses of a coin required to achieve 'Heads' for the first time,

(d) the number of cards drawn from a pack of 52 until the first heart appears.

Describe in as much detail as you can the probability space induced by each.

2. In Example 4 the random variable X denoted the sum of the numbers shown on two dice. Find the distribution of X, assuming the dice to be fair. Construct an induced probability space $(\Omega', \mathscr{F}', P')$ with $\Omega' = \{j \mid j = 2, 3, \ldots, 12\}$.

3. Two fair dice are rolled. Let X denote the number of even faces shown, and Y the number of odd. Show that X and Y are identically distributed.

4. Let $F(x)$ be the distribution function of a random variable X. For arbitrarily small δ express in terms of individual probabilities or density functions the probabilities

$$\Pr(x < X \leqslant x+\delta), \qquad \Pr(x-\delta \leqslant X < x),$$
$$\Pr(x-\tfrac{1}{2}\delta < X < x+\tfrac{1}{2}\delta), \qquad \Pr(x \leqslant X < x+\delta),$$

when

(a) X is lattice
(b) X is absolutely continuous.

5. A penny is repeatedly tossed. What is the sample space corresponding to the first n tosses? Let $I_j = 1$ if heads occur on the jth toss, $I_j = 0$ if tails. Why is I_j a random variable? What is its distribution?

6. (Continuation.) Let $n \to \infty$. What is the sample space, and is I_j still a random variable?

7. Consider the negative exponential distribution defined in Example 7. Let $A = \{(0, 2\lambda^{-1})\}$ and $B = \{(\lambda^{-1}, 3\lambda^{-1})\}$ be two elements of the induced σ field. Find $P(A \cup B)$, $P(A)$, $P(B)$, $P(AB)$, and $P(A\overline{B})$.

8. Let $f(x)$ be a density function and $(\Omega, \mathscr{F}, P)$ the induced probability space. If $A \in \mathscr{F}$, verify that

$$P(A) = \int_{x \in A} f(x)\, dx.$$

9. The distribution of X is

$$\Pr(X = x) = \binom{n}{x} p^x (1-p)^{n-x}, \qquad x = 0, 1, \ldots, n.$$

Let A and B be the events

$$A = \{\omega \mid X(\omega) \leqslant 2\}, \qquad B = \{\omega \mid X(\omega) \geqslant n-2\}.$$

Under what circumstances is $P(A) = P(B)$?

10. Suppose X is a discrete random variable with distribution $\Pr(X = x) = p_x$, $x = 0, 1, \ldots$. For fixed z in $(0, 1)$ show that $Y = z^X$ is also a random variable.

11. Let $(\Omega, \mathscr{F}, P)$ be a probability space with $\Omega = \{\omega \mid 0 < \omega < \infty\}$. Is $X = \log \omega$ a random variable? If so, find its distribution function when P is defined by $P(\{(-\infty, x]\}) = 1 - e^{-\lambda x}$ if $x > 0$, and zero otherwise.

1.5 EXPECTATION AND MOMENTS

Let X be a random variable with distribution function

$$\Pr(X \leqslant x) = F(x).$$

As stated in the last section attention is restricted to the two cases when

(a) $F(x)$ is a step function,

$$F(x) = \sum_{y \leqslant x} p_y \qquad \text{where } p_y = \Pr(X = y),$$

(b) $F(x)$ is differentiable with density $f(x)$,

$$F(x) = \int_{-\infty}^{x} f(y)\, dy.$$

Definition 1.5.1 The *expectation of* X, written EX, is defined to be

$$EX = \begin{cases} \sum_x x p_x, & \text{(discrete case)} \\ \int_{-\infty}^{\infty} x f(x)\, dx, & \text{(differentiable case)} \end{cases}$$

provided the sum or integral is absolutely convergent.* □

EX is also called the expected value of X, the mean of X, the first moment of X about the origin, or more colloquially the average of X. The precise sense in which EX is the average value of X will become clear after studying the laws of large numbers in later chapters.

Example 1 Suppose $\Omega = \{0, 1, 2, \ldots, N-1\}$ and the outcomes are equally likely. If $X(\omega) = \omega$ then

$$EX = \sum_{j=0}^{N-1} j p_j = \frac{1}{N} \sum_{j=0}^{N-1} j = \frac{N-1}{2},$$

which is the arithmetic mean of the N numbers comprising the sample space. From this point of view EX can be thought of as an arithmetic mean weighted by the corresponding probabilities. □

Example 2 Let $\Pr(X = x) = e^{-\lambda}\lambda^x/x!$, $x = 0, 1, 2, \ldots$; $\lambda > 0$, (in which case we say that X satisfies the *Poisson distribution*.)

$$EX = \sum_{x=1}^{\infty} x e^{-\lambda} \frac{\lambda^x}{x!} = \lambda e^{-\lambda} \sum_{y=0}^{\infty} \frac{\lambda^y}{y!} = \lambda.$$

The Poisson distribution is therefore uniquely determined by its mean. □

Example 3 Suppose X has the density

$$f(x) = \frac{1}{(2\pi)^{\frac{1}{2}}} e^{-\frac{1}{2}(x-\mu)^2}, \qquad -\infty < x < \infty.$$

* The general theory treats also singular distribution functions which are continuous but not differentiable and the expectation EX is defined by the Lebesgue–Stieltjes integral $EX = \int_{-\infty}^{\infty} x\, dF(x)$.

[Assume for the moment that $\int_{-\infty}^{\infty} f(x)\,dx = 1$. This is shown in § 2.5]

Then

$$EX = \frac{1}{(2\pi)^{\frac{1}{2}}}\int_{-\infty}^{\infty} xe^{-\frac{1}{2}(x-\mu)^2}\,dx$$

$$= \frac{1}{(2\pi)^{\frac{1}{2}}}\int_{-\infty}^{\infty} (\mu+y)e^{-\frac{1}{2}y^2}\,dy$$

$$= \mu+\frac{1}{(2\pi)^{\frac{1}{2}}}\int_{-\infty}^{\infty} ye^{-\frac{1}{2}y^2}\,dy = \mu. \qquad \square$$

Example 4 $\Pr(X = x) = 6/(\pi^2 x^2), \quad x = 1,2,3,\ldots,$

$$EX = \sum_{x=1}^{\infty} \frac{6}{\pi^2 x} = \infty.$$

According to Definition 1.5.1, the expectation of this random variable is not defined. $\square$

The last example illustrates that one cannot take the existence of a finite expectation for granted, although this is the case for the most commonly occurring probability distributions. Random variables with a divergent sum or integral defining EX arise naturally in the theory of Markov chains and random walks, and examples appear in § 4.5.

The expectation EX may also fail to exist in another way. For example, suppose the density $f(x)$ is symmetric, $f(x) = f(-x)$, and we define EX by

$$EX = \lim_{a\to\infty}\int_{-a}^{0} xf(x)\,dx + \lim_{b\to\infty}\int_{0}^{b} xf(x)\,dx.$$

By symmetry

$$\int_{-a}^{0} xf(x)\,dx = -\int_{0}^{a} xf(x)\,dx$$

and in the limit one obtains the indeterminate form $EX = \infty-\infty$ if $\lim_{a\to\infty} \int_0^a xf(x)\,dx = \infty$. It is to avoid such difficulties that Definition 1.5.1 requires the *absolute convergence* of the appropriate integral or sum.

Definition 1.5.1 for EX can be extended to define the expectation of a function of the random variable X.

Definition 1.5.2 The expectation of $g(X)$ is defined to be

$$Eg(X) = \begin{cases} \sum_x g(x)p_x, & \text{(discrete case),} \\ \int_{-\infty}^{\infty} g(x)f(x)\,dx, & \text{(differentiable case).} \end{cases} \tag{1.5.1}$$

provided the sum or integral converges absolutely. □

Here p_x, $f(x)$ are the probability mass function or density function of X as the case may be. From (1.5.1) it follows immediately that expectation is a linear operation, since for constants a, b,

$$\begin{aligned} E(aX+b) &= \int_{-\infty}^{\infty} (ax+b)f(x)\,dx \\ &= a\int_{-\infty}^{\infty} xf(x)\,dx + b\int_{-\infty}^{\infty} f(x)\,dx \\ &= aEX+b. \end{aligned} \tag{1.5.2}$$

An important special case is when $g(X) = (X-EX)^2$. The quantity

$$E(X-EX)^2 = \begin{cases} \sum_x (x-EX)^2 p_x & \text{(discrete case)} \\ \int_{-\infty}^{\infty} (x-EX)^2 f(x)\,dx & \text{(differentiable case)} \end{cases} \tag{1.5.3}$$

is called the *variance of* X, or the *second moment of* X *about its mean*. Just as the expectation EX measures a point of location (centre of gravity) of the distribution of X, so the variance describes the spread or dispersion of the distribution about this central point. The mean and variance are useful descriptive quantities in the sense that they summarise in numerical form information about the shape of a distribution.

By way of illustration suppose it is required to obtain bounds on the probability of a random variable realising very small or very large values relative to the mean. The celebrated result known as *Chebyshev's Inequality*

$$\Pr(|X-\mu| \geqslant \delta\sigma) \leqslant 1/\delta^2 \tag{1.5.4}$$

holds for any random variable X with mean $\mu = EX$, variance

$\sigma^2 = E(X-\mu)^2$, and for any $\delta > 0$. To prove (1.5.4) suppose to be definite that X has a density. Then

$$\begin{aligned}\sigma^2 &= \int_{-\infty}^{\infty} (x-\mu)^2 f(x)\,dx \\ &\geqslant \int_{-\infty}^{\mu-\delta\sigma} (x-\mu)^2 f(x)\,dx + \int_{\mu+\delta\sigma}^{\infty} (x-\mu)^2 f(x)\,dx \\ &\geqslant (\delta\sigma)^2 \int_{-\infty}^{\mu-\delta\sigma} f(x)\,dx + (\delta\sigma)^2 \int_{\mu+\delta\sigma}^{\infty} f(x)\,dx \\ &= (\delta\sigma)^2 \Pr(|X-\mu| \geqslant \delta\sigma).\end{aligned}$$

Hence (1.5.4).

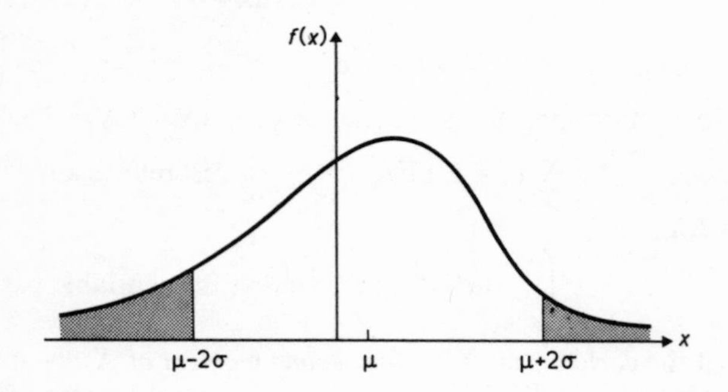

For example, take $\delta = 2$. The probability $\Pr(|X-\mu| \geqslant 2\sigma)$ is the shaded area under the curve $f(x)$ for $|x-\mu| \geqslant 2\sigma$. From (1.5.4) it follows that this area is $\leqslant 2^{-2} = 0{\cdot}25$ irrespective of the functional form of $f(x)$ provided only that X has a mean and variance. If $f(x)$ is known then in general one can sharpen this result, but there do exist random variables for which the bound in (1.5.4) is attained. (See Exercises 5, 6).

The notation $\mu = EX$ and $\sigma^2 = E(X-EX)^2 = E(X-\mu)^2$ for the mean and variance is widely used. The positive square root σ of the variance is called the *standard deviation* of X. Frequently we write $\sigma^2 = \text{Var}\, X$.

A useful formula which we prove for the continuous case is

$$\begin{aligned}\sigma^2 &= \int_{-\infty}^{\infty} (x-\mu)^2 f(x)\,dx \\ &= \int_{-\infty}^{\infty} (x^2-2\mu x+\mu^2) f(x)\,dx \\ &= \int_{-\infty}^{\infty} x^2 f(x)\,dx - 2\mu \int_{-\infty}^{\infty} x\,f(x)\,dx + \mu^2 \\ &= EX^2-\mu^2. \end{aligned} \tag{1.5.5}$$

EX^2 is called the *second moment of X about the origin.* More generally, if b is any constant the *rth moment of X about b* is defined to be $E(X-b)^r$, provided the defining sum or integral is absolutely convergent. Using the binomial theorem to expand $(X-b)^r$ we find that the rth moment about b is related to the first r moments about the origin by

$$E(X-b)^r = \sum_{j=0}^{r} \binom{r}{j} (EX^j)(-b)^{r-j}$$

(1.5.5) is the special case $r = 2$, $b = \mu$.

Recall that the defining integral or sum must be absolutely convergent before we agree that a moment exists. In particular

$$\begin{aligned} EX^r &= \int_{-\infty}^{\infty} x^r f(x)\,dx \\ &= \int_{-\infty}^{\infty} x^r f(x)\,dx + \int_{-\infty}^{0} x^r f(x)\,dx \end{aligned}$$

and hence both $\lim_{T_1\to\infty} \int_0^{T_1} x^r f(x)\,dx$ and $\lim_{T_2\to\infty} \int_{-T_2}^{0} x^r f(x)\,dx$ are required to converge to finite limits for EX^r to be defined. It is possible to obtain a useful alternative expression for EX^r.

An integration by parts yields for all $r > 0$

$$\int_0^{T_1} x^r f(x)\,dx = -T_1^r(1-F(T_1)) + \int_0^{T_1} r x^{r-1}(1-F(x))\,dx$$

$$\int_{-T_2}^{0} x^r f(x)\,dx = -(-T_2)^r F(-T_2) - \int_{-T_2}^{0} r x^{r-1} F(x)\,dx. \tag{1.5.6}$$

For the time being consider only the first identity in (1.5.6). If EX^r exists then $\int_0^\infty x^r f(x)\,dx < \infty$ and

$$\lim_{T_1\to\infty}\int_{T_1}^{\infty} x^r f(x)\,dx = 0.$$

But

$$\int_{T_1}^{\infty} x^r f(x)\,dx \geqslant T_1^r \int_{T_1}^{\infty} f(x)\,dx = T_1^r(1-F(T_1)).$$

Hence the existence of EX^r implies

$$\int_0^\infty x^r f(x)\,dx = \int_0^\infty rx^{r-1}(1-F(x))\,dx.$$

Conversely, suppose $\int_0^\infty rx^{r-1}(1-F(x))\,dx < \infty$. Then

$$\lim_{T_1\to\infty}\int_0^{T_1} x^r f(x)\,dx < \infty$$

since, from (1.5.6),

$$\int_0^{T_1} x^r f(x)\,dx \leqslant \int_0^{T_1} rx^{r-1}(1-F(x))\,dx.$$

Precisely the same argument applies to the second member of (1.5.6) and, recognising that $\Pr(|X| > x) = 1-F(x)+F(-x)$, we have

Theorem 1.5.1 A necessary and sufficient condition for EX^r to exist is that $x^{r-1}\Pr(|X| > x)$ be absolutely integrable. In that case

$$\begin{aligned} EX^r &= \int_{-\infty}^{\infty} x^r f(x)\,dx \\ &= \int_0^\infty rx^{r-1}(1-F(x))\,dx - \int_{-\infty}^{0} rx^{r-1}F(x)\,dx. \qquad \square \end{aligned}$$

The theorem clarifies to an extent the way in which moments depend on the amount of probability in the tails of a distribution.

Example 5 The *Cauchy distribution* is defined by the density

$$f(x) = \frac{1}{\pi(1+x^2)}, \qquad -\infty < x < \infty.$$

A trigonometric transformation shows that $\int_{-\infty}^{\infty} f(x)\,dx = 1$. However the mean EX is not defined since

$$\int_0^T \frac{x}{\pi(1+x^2)}\,dx = \frac{1}{2\pi}\log(1+x^2)\Big|_0^T$$
$$\to \infty \quad \text{as } T \to \infty.$$

Also $\Pr(|X| > x) \sim cx^{-1}$ as $x \to \infty$ and x^{-1} is not integrable over $(0, \infty)$. □

There are other useful choices of the function g in Definition 1.5.2 apart from $g(x) = x^r$. Three such functions are

$$g(x) = z^x, \qquad g(x) = e^{-sx}, \qquad g(x) = e^{isx},$$

where in each case z or s is a 'dummy' variable constrained to take real values which ensure that $Eg(X)$ is defined.

Suppose firstly that X is lattice, distributed over the non-negative integers with

$$\Pr(X = x) = p_x, \qquad x = 0, 1, 2, \ldots.$$

The power series

$$G(z) = Ez^X = \sum_{x=0}^{\infty} z^x p_x, \tag{1.5.7}$$

corresponding to the choice $g(x) = z^x$, is called the *probability generating function of* X or the *generating function of* $\{p_x\}$. $G(z)$ is absolutely convergent for at least $|z| \leqslant 1$ since for those values of z

$$|Ez^X| \leqslant \sum_{x=0}^{\infty} p_x = 1.$$

Generating functions are discussed further in § 2.2. For the present we note only their use to obtain moments of X. Suppose $\mu = EX$ and $\sigma^2 = \text{Var}\, X$ are finite. A formal procedure to the limit yields

$$G'(1) = \sum_{x=0}^{\infty} x p_x,$$

$$G''(1) = \sum_{x=0}^{\infty} x(x-1)p_x = EX^2 - EX,$$

where $G'(1)$ denotes $\lim_{z \to 1-} G'(z)$. Hence μ and σ^2 can be found from $G(z) = Ez^X$ by

$$\mu = G'(1), \qquad \sigma^2 = G''(1) + G'(1) - (G'(1))^2.$$

Example 6 $p_x = (1-\rho)\rho^x$, $x = 0, 1, 2, \ldots; 0 < \rho < 1$.

$$G(z) = \sum_{x=0}^{\infty} (1-\rho)(z\rho)^x = \frac{1-\rho}{1-\rho z}$$

which is convergent for $|z| < \rho^{-1}$. Since $G'(1) = \rho(1-\rho)^{-1}$, $G''(1) = 2\rho^2(1-\rho)^{-2}$ we find $\mu = \rho(1-\rho)^{-1}$ and $\sigma^2 = \rho(1-\rho)^{-2}$. □

Secondly suppose that for s in some non-degenerate interval the quantities

$$Ee^{-sX} = \begin{cases} \sum_x e^{-sX} p_x & \text{(discrete case)} \\ \int_{-\infty}^{\infty} e^{-sx} f(x)\, dx & \text{(differentiable case)} \end{cases} \tag{1.5.8}$$

are convergent. Ee^{-sX} is called* the *moment generating function of X*. The reason for the name is that the expansion of Ee^{-sX} as a power

* Frequently Ee^{sX} is defined to be the moment generating function of X. Our notation conforms with that usually used in the theory of Laplace transforms.

series in s gives

$$Ee^{-sX} = \int_{-\infty}^{\infty} \left\{ \sum_{j=0}^{\infty} (-s)^j x^j / j! \right\} f(x)\, dx$$

$$= \sum_{j=0}^{\infty} \{(-s)^j / j!\} \int_{-\infty}^{\infty} x^j f(x)\, dx$$

provided the interchange of summation and integration is valid. EX^j thus appears as the coefficient of $(-s)^j/j!$ in the expansion of Ee^{-sX} and

$$(-1)^j \frac{d^j}{ds^j}(Ee^{-sX})\bigg|_{s=0} = EX^j.$$

Example 7 Let X have the negative exponential density

$$\begin{aligned} f(x) &= \lambda e^{-\lambda x}, \quad x \geqslant 0, \\ &= 0 \qquad\quad\; x < 0, \end{aligned}$$

where $\lambda > 0$. Then

$$Ee^{-sX} = \int_0^{\infty} \lambda e^{-(\lambda+s)x}\, dx = \frac{-\lambda e^{-(\lambda+s)x}}{\lambda+s}\bigg|_0^{\infty}$$

$$= \begin{cases} \infty & \text{if } s \leqslant -\lambda, \\ \lambda(\lambda+s)^{-1} & \text{if } s > -\lambda. \end{cases}$$

Ee^{-sX} is defined only for s in $(-\lambda, \infty)$. □

Our third choice of g, namely $g(x) = e^{isx}$, with $i = \sqrt{-1}$, leads to a quantity of considerable importance in probability theory. For s real

$$Ee^{isX} = \begin{cases} \sum_x e^{isx} p_x & \text{(discrete case)} \\ \int_{-\infty}^{\infty} e^{isx} f(x)\, dx & \text{(differentiable case)} \end{cases} \tag{1.5.9}$$

is called the *characteristic function of* X. It is clearly related to the moment generating function just introduced (indeed EX^j is the coefficient of $(is)^j/j!$ in the series expansion of Ee^{isX}) but the characteristic function has the special advantage that it is defined for all real s for all random variables. To see this suppose X is continuous and write

$$\begin{aligned} Ee^{isX} &= E(\cos sX + i \sin sX) \\ &= \int_{-\infty}^{\infty} \cos sx\, f(x)\, dx + i \int_{-\infty}^{\infty} \sin sx\, f(x)\, dx. \end{aligned}$$

Both $\cos sx$ and $\sin sx$ are dominated by unity in absolute value, implying that the two real integrals comprising Ee^{isX} are absolutely convergent for all s in $(-\infty, \infty)$.

Probability generating functions, moment generating functions, and characteristic functions will be collectively referred to as *generating functions* of random variables, or more correctly as generating functions of the distributions of random variables. They constitute an important analytical tool and are discussed in some detail in the next chapter.

Exercises

1. Without evaluating sums or integrals show that

$$E(\sum_{j=1}^{n} a_j X^j) = \sum_{j=1}^{n} a_j(EX^j)$$

$$E(\sum_{j=1}^{n} a_j X^j)^2 = \sum_{j=1}^{n} a_j^2(EX^{2j}) + 2\sum_{j=1}^{k-1} \sum_{k=2}^{n} a_j a_k (EX^{j+k}).$$

2. The distribution of X is given by $\Pr(X = 1) = p$, $\Pr(X = 2) = q$ with $p+q = 1$. Suppose $Y = X^2$. Find EY by using (a) (1.5.1), (b) the formula $EY = \sum_y y \Pr(Y = y)$. Also find the variance of Y.

3. If X is a random variable on $(-\infty, \infty)$ show that $X^+ = \max(0, X)$ and $X^- = \max(0, -X)$ are also random variables. Write down the moment generating function of X^+, X^-, and verify that

$$EX = EX^+ - EX^-, \quad E|X| = EX^+ + EX^-.$$

4. Suppose a, b are constants and X is a random variable with moment generating function $M(s)$. Extend (1.5.2) to show that $\text{Var}(aX+b) = a^2\text{Var}\, X$ and $Ee^{-s(aX+b)} = e^{-bs}M(as)$.

5. Let X be distributed as

$$\Pr(X = -1) = \tfrac{1}{8}, \quad \Pr(X = 0) = \tfrac{3}{4}, \quad \Pr(X = 1) = \tfrac{1}{8}.$$

If μ, σ^2 are respectively the mean and variance of X, show that equality in fact holds in the Chebyshev inequality (1.5.4) when $\delta = 2$.

6. Compare the bound given by Chebyshev inequality with the exact results for $\Pr(|X-\mu|) \geqslant 2\sigma$ for distributions defined by

$$\text{(a) } f(x) = \frac{1}{3(2\pi)^{\frac{1}{2}}} \exp\left\{-\frac{(x-10)^2}{18}\right\}, -\infty < x < \infty,$$

$$\text{(b) } f(x) = 2e^{-2x}, \quad 0 < x.$$

7. Show that $E|X|^r > E|X|^j$ if $r > j$ and X can take values outside the unit interval. It follows that the existence of an rth moment implies the existence of all moments of lower order.

8. Since the series $c(\alpha) = \sum_{j=1}^{\infty} j^{-(1+\alpha)}$ is convergent for $\alpha > 0$, it is possible to define a random variable X_α whose distribution is

$$\Pr(X_\alpha = j) = \frac{1}{c(\alpha) j^{1+\alpha}}, \qquad j = 1, 2, 3, \ldots$$

For what values of α does X have a mean, variance, and kth moment? Check that the same answer holds for the continuous random variable Y_α with density

$$f(x) = \frac{1}{y^{1+\alpha} \int_1^\infty y^{-(1+\alpha)}\, dy}, \qquad 1 < y < \infty.$$

9. Find the moment generating functions (and characteristic functions) corresponding to the densities

$$\text{(a)} \quad f(x) = \frac{1}{\sigma(2\pi)^{\frac{1}{2}}} \exp\left[-\frac{(x-\mu)^2}{2\sigma^2}\right], \qquad -\infty < x < \infty,$$

(b) $f(x) = e^{-\lambda x}x\lambda^2, \qquad 0 < x < \infty,$

(c) $f(x) = 1, \qquad 0 < x < 1.$

Show that in (a) the mean is μ and the variance σ^2.

10. Find the generating functions of

(a) $p_j = 1/m, \qquad j = 0, 1, 2, \ldots, m-1,$

(b) $p_j = \binom{N}{j} p^j(1-p)^{N-j}, \qquad j = 0, 1, \ldots, N; 0 < p < 1,$

(c) $p_j = e^{-\lambda}\lambda^j/j!, \qquad j = 0, 1, \ldots; \lambda > 0$

11. Two fair coins are tossed. A gambler wins two dollars if both coins show heads and otherwise loses a dollar. What is his expected gain? Would you call the game fair?

12. Use tables to sketch the density function

$$f(x) = \frac{1}{\sigma(2\pi)^{\frac{1}{2}}}\exp\left[\frac{-x^2}{2\sigma^2}\right], \qquad -\infty < x < \infty,$$

for $\sigma = 1, 2, 3, 5$. Note that the curves become flatter as σ increases, indicating a probability 'flow' away from the mean with increasing variance

13. The mean $\mu = EX$ is a measure of location or central tendency of the distribution function $\Pr(X \leqslant x)$. Two other measures of location are

(a) the *median* $\tilde{x}$, defined by $\Pr(X \leqslant \tilde{x}) = \frac{1}{2} = \Pr(X > \tilde{x})$,
(b) the *mode* x_m, defined to be that value of x which maximises $f(x)$ or $\Pr(X = x)$.

Note that the mode need not be unique (e.g. the density in Exercise 1.2.14b is bimodal). Show that if

$$f(x) = \frac{1}{\sigma(2\pi)^{\frac{1}{2}}}\exp\left[-\frac{(x-\mu)^2}{2\sigma^2}\right], \qquad -\infty < x < \infty,$$

then mean, median and mode are all equal to μ, but for

$$f(x) = \lambda e^{-\lambda x}, \qquad 0 < x,$$

the three quantities are distinct.

14. Let X have density $f(x)$. Show that of all second moments $E(X-b)^2$ the variance is the smallest, and that $E|X-b|$ is minimised when b is the median $\tilde{x}$.

15. (Continuation of Exercise 1.4.10.) For fixed z in (0, 1) and X distributed on the non-negative integers, $Y = z^X$ is a random variable on $1, z, z^2, \ldots$, with expected value $EY = Ez^X = G(z)$. Show that $\operatorname{Var} Y = G(z^2) - (G(z))^2$, which incidentally implies $G(z^2) > (G(z))^2$, $0 < z < 1$, for all probability generating functions. What is the corresponding result if X is continuous and has a moment generating function?

APPENDIX

Monotone sequences of events

Suppose $A_1, A_2, A_3, \ldots$ are events on $(\Omega, \mathscr{F}, P)$ which are non-increasing,

$$A_1 \supseteq A_2 \supseteq A_3 \supseteq \ldots.$$

The limit of the non-increasing sequence $\{A_n, n = 1, 2, \ldots\}$ is defined to be

$$A = \lim_{n\to\infty} A_n = \bigcap_{n=1}^{\infty} A_n. \qquad \text{(A1.1)}$$

Since $\mathscr{F}$ is a σ field, the limit event A is a member of $\mathscr{F}$.

Similarly if $B_1, B_2, B_3, \ldots$ are non-decreasing events on $(\Omega, \mathscr{F}, P)$,

$$B_1 \subseteq B_2 \subseteq B_3 \subseteq \ldots$$

the limit of $\{B_n, n = 1, 2, \ldots\}$ is defined by

$$B = \lim_{n\to\infty} B_n = \bigcup_{n=1}^{\infty} B_n. \qquad \text{(A1.2)}$$

The notion of a limit is thus defined by (A1.1) or (A1.2) for all monotone sequences of events. As an example let $\Omega = (-\infty, \infty)$ and

$$A_n = \left\{x \mid 1 - \frac{1}{n} < x < 2 + \frac{1}{n}\right\}, \qquad n = 1, 2, \ldots.$$

Then

$$A_1 \supseteq A_2 \supseteq A_3 \supseteq \ldots$$

and

$$\lim_{n\to\infty} A_n = \{x \mid 1 \leqslant x \leqslant 2\}.$$

Monotone sequences of events enjoy the following important property;

Theorem If $\{A_n, n = 1, 2, \ldots\}$ is a monotone sequence of events on $(\Omega, \mathscr{F}, P)$ then

$$\lim_{n\to\infty} P(A_n) = P(\lim_{n\to\infty} A_n).$$

Proof Consider first the case when the A_n are non-increasing. If

$$A = \bigcap_{n=1}^{\infty} A_n$$

we have to show that

$$\lim_{n \to \infty} P(A_n) = P(A).$$

Decompose A_n into the union of disjoint events as follows

$$A_n = \bigcup_{j=n}^{\infty} A_j \bar{A}_{j+1} \cup A.$$

Note that $AA_j \bar{A}_{j+1} = \varnothing$ for every $j = 1, 2, \ldots$ since $A = \bigcap_{n=1}^{\infty} A_n$, and hence $A \bigcup_{j=n}^{\infty} A_j \bar{A}_{j+1} = \varnothing$. Then by the countable additivity of P

$$P(A_n) = \sum_{j=n}^{\infty} P(A_j \bar{A}_{j+1}) + P(A).$$

But

$$\sum_{i=1}^{\infty} P(A_j \bar{A}_{j+1}) \leqslant P(\Omega) = 1$$

implies

$$\lim_{n \to \infty} \sum_{j=n}^{\infty} P(A_j \bar{A}_{j+1}) = 0.$$

Therefore

$$\lim_{n \to \infty} P(A_n) = P(A),$$

which proves the theorem for non-increasing A_n.

To avoid confusion suppose now that $\{B_n\}$ is a non-decreasing sequence with

$$\lim_{n \to \infty} B_n = \bigcup_{n=1}^{\infty} B_n = B.$$

To show that $\lim_{n \to \infty} P(B_n) = P(B)$, consider the sequence of complementary events $\{\bar{B}_n\}$. Since the B_n are non-decreasing the $\bar{B}_n$ must be non-increasing, and by the first part of the proof,

$$\lim_{n \to \infty} P(\bar{B}_n) = P\left(\bigcap_{n=1}^{\infty} \bar{B}_n\right).$$

But

$$\left(\bigcap_{n=1}^{\infty}\bar{B}_n\right)\cup\left(\bigcup_{n=1}^{\infty}B_n\right) = \Omega,$$

and since the events on the left-hand side are disjoint,

$$\begin{aligned}
P\left(\bigcup_{n=1}^{\infty}B_n\right) &= 1-P\left(\bigcap_{n=1}^{\infty}\bar{B}_n\right)\\
&= 1-\lim_{n\to\infty} P(\bar{B}_n)\\
&= 1-\lim_{n\to\infty}\{1-P(B_n)\}\\
&= \lim_{n\to\infty} P(B_n).
\end{aligned}$$

This completes the proof of the theorem. □

The limits of more general sequences of events are considered in Chapter 5.

Chapter 2
SOME REAL VARIABLE THEORY

Later chapters are essentially concerned with probabilistic questions stated in terms of random variables or sequences of random variables. Certain mathematical techniques are particularly useful, and Chapter 2 is designed to collect together for reference results of a primarily analytical nature. The first three sections treat generating functions. § 4 is traditional calculus dealing with change of variable methods and their interpretation in probabilistic terms. As its title indicates § 5 deals mainly with special functions which arise in connection with commonly occurring distributions, although indicator functions do have a wider significance which we do not develop.

The chapter need not be read systematically. Readers not particularly concerned with the mathematical background (or those who already have it) could omit much at least of §§ 2.1–3 provided they appreciate

(a) the 1–1 relationship between distribution functions and generating functions (stated in Theorems 2.2.1 (c), 2.3.1),
(b) moments can be found as the coefficients in appropriate Taylor expansions (Theorems 2.2.3, 2.3.5, 2.3.9),
(c) generating functions translate convolutions into products (Theorems 2.2.4, 2.3.2, Example 2.3.2).

Mainly because of (a)–(c) above, the generating function has proved a useful tool in probability theory. By virtue of (a), problems can be posed (and their answers attempted) in terms of the function that is easiest to handle, and this fact is extensively exploited in Chapter 4. The interpretation of convolution as it arises in probability theory is deferred until Chapter 3. For the moment we treat it as a purely mathematical operation.

2.1 TAYLOR'S THEOREM

The aim of this section is to apply a version of Taylor's Theorem to the generating functions introduced in § 1.5. First some notation.

Definition 2.1.1 *O and o symbols.* For given functions $h(x) > 0$ and $g(x)$ the notation

$$g(x) = O(h(x)), \qquad x \to b,$$

means $|g(x)| \leqslant c\,h(x)$ when x is arbitrarily close to b, with c some positive constant, and

$$g(x) = o(h(x)), \qquad x \to b,$$

means $\lim_{x \to b} \{g(x)/h(x)\} = 0.$ □

In particular $g(x) = O(1)$ signifies that $g(x)$ is bounded as $x \to b$ and $g(x) = o(1)$ is another way of stating $\lim_{x \to b} g(x) = 0$. For example, continuity of $g(x)$ at $x = b$ can be expressed by

$$g(x) = g(b) + o(1), \qquad x \to b.$$

Again, if $g(x)$ is, say, differentiable from the right at the point y, in which case

$$\lim_{\delta \to 0} \frac{g(y+\delta)-g(y)}{\delta} = g'(y),$$

then

$$g(y+\delta) = g(y) + \delta\, g'(y) + o(\delta), \qquad \delta \to 0. \tag{2.1.1}$$

This expression was used in § 1.2. Note that

$$o(h(x)) = h(x)o(1).$$

A related notation is

$$g(x) \sim h(x), \qquad x \to b,$$

to mean

$$\lim_{x \to b} g(x)/h(x) = 1.$$

For example, $\{x/(1+x^2)\} = O(x^{-1})$ as $x \to \infty$. Indeed

$$\{x/(1+x^2)\} \sim x^{-1}, x \to \infty$$

However, as $x \to 0$, $\{x/(1+x^2)\} = o(1)$.

Taylor's theorem expresses the value of a 'well behaved' function at a particular point in terms of the function and its derivatives evaluated at another point in its domain of definition. Proofs are readily available, but one is supplied because of our frequent use of the result and, besides, the argument is brief.

Theorem 2.1.1 (Taylor) Suppose $g(x)$ has $r-1$ continuous derivatives on $[a,b]$ and that the rth derivative $g^{(r)}(x)$ is defined at least on the open interval (a,b). Then for any x, $a < x < b$, there is a ξ_0 in (x, b) and a ξ_1 in (a, x) such that

$$g(x) = g(b)+\sum_{j=1}^{r-1}\frac{(x-b)^j}{j!}g^{(j)}(b)+\frac{(x-b)^r}{r!}g^{(r)}(\xi_0), \quad (2.1.2)$$

and

$$g(x) = g(a)+\sum_{j=1}^{r-1}\frac{(x-a)^j}{j!}g^{(j)}(a)+\frac{(x-a)^r}{r!}g^{(r)}(\xi_1). \quad (2.1.3)$$

Proof We prove only (2.1.2) as (2.1.3) is established in the same way. Taking x fixed in (a, b), define k by

$$g(x) = g(b)+\sum_{j=1}^{r-1}\frac{(x-b)^j}{j!}g^{(j)}(b)+\frac{(x-b)^r}{r!}k,$$

and, for $x \leqslant y \leqslant b$, let

$$R(y) = -g(x)+g(y)+\sum_{j=1}^{r-1}\frac{(x-y)^j}{j!}g^{(j)}(y)+\frac{(x-y)^r}{r!}k.$$

Then

$$R(x) = 0 = R(b)$$

and, if $a < y < b$,

$$\begin{aligned}\frac{dR(y)}{dy} &= g'(y)+\sum_{j=1}^{r-1}\left[\frac{(x-y)^j}{j!}g^{(j+1)}(y)-\frac{(x-y)^{j-1}}{(j-1)!}g^{(j)}(y)\right] \\ &\quad -\frac{(x-y)^{r-1}}{(r-1)!}k \\ &= \frac{(x-y)^{r-1}}{(r-1)!}\{g^{(r)}(y)-k\}.\end{aligned}$$

Continuity of $R(y)$ on $[x, b]$ and its vanishing at the two end points of the interval imply that there is a point ξ_0, $x < \xi_0 < b$, such that $R'(\xi_0) = 0$. But this is the same as $k = g^{(r)}(\xi_0)$, and (2.1.2) is proved. □

If the rth derivate of $g(x)$ exists (perhaps only as a left or right-hand limit) at the end points a and b, it is possible to express (2.1.2) and (2.1.3) in a different way. Suppose firstly that $g^{(r)}(x)$ is continuous on $(b-\varepsilon, b]$ for some $\varepsilon > 0$. That is, suppose

$$g^{(r)}(b) = \lim_{\delta \to 0+} g^{(r)}(b-\delta)$$

exists*. Since $x < \xi_0 < b$ it follows that as $x \to b$

$$g^{(r)}(\xi_0) = g^{(r)}(b) + o(1),$$
$$(x-b)^r g^{(r)}(\xi_0) = (x-b)^r g^{(r)}(b) + o((b-x)^r),$$

and (2.1.2) can therefore be rewritten as

$$g(x) = g(b) + \sum_{j=1}^{r} \frac{(x-b)^j}{j!} g^{(j)}(b) + o((b-x)^r), \qquad x \to b. \qquad (2.1.4)$$

In the same way the assumption that $g^{(r)}(x)$ is continuous from the right at $x = a$ implies from (2.1.3) that

$$g(x) = g(a) + \sum_{j=1}^{r} \frac{(x-a)^j}{j!} g^{(j)}(a) + o((x-a)^r), \qquad x \to a. \qquad (2.1.5)$$

Before specialising the discussion to generating functions we state, for ease of reference, some standard results on the interchange of limit operations.

Theorem 2.1.2 If $v_n(x)$ has a continuous derivative $v_n'(x)$, $n = 0, 1, 2, \ldots$, and if $\sum_{n=0}^{\infty} v_n'(x)$ is uniformly convergent for x in an interval

* The notation $\lim_{\delta \to 0+}$ means that δ approaches zero through positive values and we could have written $\lim_{\delta \downarrow 0}$. Similarly $\lim_{\delta \to c-}$ signifies that the limit is taken from below, also expressed by $\lim_{\delta \uparrow c}$. Frequently we will write simply $\lim_{\delta \to c}$, the set through which the limit is taken being clear from the context.

(a,b), then $g(x) = \sum_{n=0}^{\infty} v_n(x)$ can be differentiated term by term and

$$\frac{d}{dx}g(x) = \sum_{n=0}^{\infty} v_n'(x), \qquad a < x < b. \qquad \square$$

Theorem 2.1.3 Let $\sum_{n=0}^{\infty} v_n(x)$ be uniformly convergent for $a < x < b$. Then

(i) $\lim_{x\to c} \sum_{n=0}^{\infty} v_n(x)\, dx = \sum_{n=0}^{\infty} v_n(c)$ provided $\lim_{x\to c} v_n(x) = v_n(c)$, where $a < c < b$,

(ii) $$\int_{x_0}^{x_1} \sum_{n=0}^{\infty} v_n(x)\, dx = \sum_{n=0}^{\infty} \int_{x_0}^{x_1} v_n(x)\, dx, a < x_0 < x_1 < b. \qquad \square$$

Useful expansions for generating functions follow from these results. If $\{p_j\}$ is a probability distribution on the non-negative integers, its generating function

$$G(z) = \sum_{j=0}^{\infty} z^j p_j$$

converges for a least $0 \leqslant z \leqslant 1$. Suppose more generally that $G(z)$ is convergent for at least $0 \leqslant z \leqslant R_0$ where $R_0 \geqslant 1$. Then it is easy to see that for each $k = 1, 2, 3, \ldots$, the series

$$\sum_{j=k}^{\infty} j(j-1)\ldots(j-k+1)z^{j-k}p_j$$

converges uniformly at least in the half-open interval $0 \leqslant z < R_0$. Writing $(j)_k = j(j-1)\ldots(j-k+1)$, a consequence of Theorem 2.1.2 is that for $k = 1, 2, 3, \ldots$,

$$G^{(k)}(z) = \sum_{j=k}^{\infty} (j)_k z^{j-k} p_j, \qquad 0 \leqslant z < R_0;$$

that is, $G(z)$ is term-wise differentiable as many times as we please at interior points of $[0, R_0)$. If it is also given that $G(z)$ is r times differentiable at $z = R_0$ then (2.1.4) applies with $b = R_0$ and we have the Taylor expansion

$$G(z) = G(R_0) + \sum_{j=1}^{r} \frac{(z-R_0)^j}{j!} G^{(j)}(R_0) + o((R_0-z)^r), \quad z \to R_0. \qquad (2.1.6)$$

If it happens that $G(z)$ in fact converges in an interval $0 \leqslant z < R$ which includes R_0 as an *interior* point the argument above shows that the derivatives $G^{(k)}(R_0)$ are well defined for all $k = 1, 2, 3, \ldots$, and the possibility arises of writing the right hand side of (2.1.6) as an infinite series. Perhaps the best known form of Taylor's Theorem asserts that this can be done for points R_0 contained inside the interval over which $G(z)$ converges. Derivatives at $R_0 = 1$ are of interest to us and we state the Taylor expansion for probability generating functions as follows:

Theorem 2.1.4 If $G(z) = \sum_{j=0}^{\infty} z^j p_j$ is defined only for $|z| \leqslant 1$ but $G^{(r)}(1) < \infty$, then (from 2.1.6))

$$G(z) = G(1) + \sum_{j=1}^{r} \frac{(z-1)^j}{j!} G^{(j)}(1) + o((1-z)^r), \qquad z \to 1-. \qquad (2.1.7)$$

On the other hand if $G(z)$ converges for z in an interval which includes $z = 1$ as an interior point then $G(z)$ is infinitely differentiable at $z = 1$, and in the neighbourhood of this point

$$G(z) = G(1) + \sum_{j=1}^{\infty} \frac{(z-1)^j}{j!} G^{(j)}(1). \qquad (2.1.8) \; \square$$

A similar result holds for the moment generating function

$$M(s) = \int_0^{\infty} e^{-sx} f(x)\, dx$$

in which $f(x)$ is a density on $[0, \infty)$. The monotonicity of e^{-sx} implies $M(s) \leqslant 1$ for $s \geqslant 0$ and we will suppose that in fact the integral defining $M(s)$ converges for $s \geqslant -\gamma_0$ with $\gamma_0 \geqslant 0$.

The argument is slightly more complicated than in the case of probability generating functions, but careful use of Theorems 2.1.2 and 2.1.3 (ii) shows that

$$M^{(k)}(s) = (-1)^k \int_0^{\infty} x^k e^{-sx} f(x)\, dx, \qquad s > -\gamma_0, k = 1, 2, 3, \ldots$$

The derivatives of interest are those at $s = 0$. If $\gamma_0 = 0$ and $M^{(r)}(0) < \infty$ then (2.1.5) applies with $a = 0$. On the other hand if $\gamma_0 > 0$ we have an infinite Taylor expansion and find, corresponding to the previous theorem:

Theorem 2.1.5 If $M(s) = \int_0^\infty e^{-sx} f(x)\,dx$ is convergent only for $s \geqslant 0$ and $M^{(r)}(0) < \infty$ then

$$M(s) = M(0) + \sum_{j=1}^{r} \frac{s^j}{j!} M^{(j)}(0) + o(s^r), \qquad s \to 0+, \qquad (2.1.9)$$

but if $M(s)$ converges for $s > -\gamma$, with $\gamma > 0$ then

$$M(s) = M(0) + \sum_{j=1}^{\infty} \frac{s^j}{j!} M^{(j)}(0). \qquad (2.1.10)$$

□

Example 1 If m is a positive integer and $c = \{\sum_{j=1}^{\infty} j^{-(m+2)}\}^{-1}$ then

$$G(z) = c \sum_{j=1}^{\infty} z^j j^{-(m+2)}$$

is a probability generating function. $G(z)$ has finite derivatives up to order m on the closed interval $[0, 1]$. Theorem 2.1.4 applies with $r = m$ but not for any $r > m$. A similar result holds for the moment generating function of the density $f(x) = (m+1)x^{-(m+2)}$, $1 \leqslant x$. □

Example 2 If $0 < \rho < 1$, $G(z) = (1-\rho)(1-\rho z)^{-1}$ is the generating function of the geometric distribution which converges for $|z| < \rho^{-1}$, (2.1.8) becomes

$$G(z) = 1 + \sum_{j=1}^{\infty} \frac{(z-1)^j j!\rho^j}{j!(1-\rho)^j}, \qquad |z-1| < (1-\rho)\rho^{-1}.$$

The moment generating function of the negative exponential distribution is

$$M(s) = \int_0^\infty e^{-sx} \lambda e^{-\lambda x}\,dx = \frac{\lambda}{\lambda + s}, \qquad s > -\lambda.$$

(2.1.10) holds for $|s| < \lambda$ with

$$M^{(j)}(0) = \frac{(-1)^j j!}{\lambda^j}.$$

□

Example 3 It is not necessarily the case that a function $g(s)$ can be expanded as in (2.1.10) even if it is infinitely differentiable at the origin. For example,

$$g(s) = \exp(-s^{-2}),$$

with $g(0) = 0$, has $g^{(j)}(0) = 0, j = 1, 2, 3, \ldots$, and the right-hand side of (2.1.10) is zero for all s. This $g(s)$ happens not to be a moment generating function. □

The third of the generating functions introduced in §1.5 was the characteristic function

$$\psi(s) = \begin{cases} \sum_x e^{isx} \Pr(X = x), & \text{(discrete case)} \\ \int_{-\infty}^{\infty} e^{isx} f(x)\, dx, & \text{(differentiable case).} \end{cases}$$

All characteristic functions can be expressed in terms of their real and imaginary parts which are ordinary real integrals (or sums),

$$\psi(s) = U(s) + iV(s) \tag{2.1.11}$$

where, in the differentiable case,

$$U(s) = \int_{-\infty}^{\infty} \cos sx\, f(x)\, dx,$$

$$V(s) = \int_{-\infty}^{\infty} \sin sx\, f(x)\, dx.$$

The preceding theory applies to $\psi(s)$ provided it holds for both the real valued functions $U(s)$ and $V(s)$. Unlike the integral defining a moment generating function, $U(s)$ and $V(s)$ are not necessarily monotonic functions and can be negative for some values of s. However, positivity was not required for the basic results (2.1.2)–(2.1.5).

In particular if $U(s)$ and $V(s)$ are r times differentiable in an interval including the origin, then the conditions of Theorem 2.1.1 are satisfied and the argument used to obtain (2.1.4) and (2.1.5) shows that as $s \to 0$

$$\begin{aligned} U(s) &= U(0) + \sum_{j=1}^{r} \frac{s^j}{j!} U^{(j)}(0) + o(s^r) \\ V(s) &= V(0) + \sum_{j=1}^{r} \frac{s^j}{j!} V^{(j)}(0) + o(s^r). \end{aligned} \tag{2.1.12}$$

But from (2.1.11) the derivatives of the complex function $\psi(s)$ are given by

$$\psi^{(j)}_{(s)} = U^{(j)}(s) + iV^{(j)}(s).$$

Adding the two parts of (2.1.12) to form $\psi(s) = U(s)+iV(s)$, we have a Taylor expansion for characteristic functions:

Theorem 2.1.6 If the characteristic function $\psi(s)$ is r times differentiable at the origin, then as $s \to 0$

$$\psi(s) = \psi(0) + \sum_{j=1}^{r} \frac{s^j}{j!}\psi^{(j)}(0) + o(s^r). \qquad \square$$

Example 4 The negative exponential distribution has characteristic function

$$\begin{aligned}\psi(s) &= \int_0^\infty \lambda e^{-(\lambda - is)x}\, ds \\ &= \frac{\lambda}{\lambda - is} = \frac{\lambda(\lambda+is)}{\lambda^2+s^2}.\end{aligned}$$

Both real and imaginary parts are continuous and infinitely differentiable for all real s, and indeed, for $|s| < \lambda$,

$$\begin{aligned}U(s) &= \frac{\lambda^2}{\lambda^2+s^2} = \sum_{j=0}^{\infty} \frac{(-1)^j s^{2j}}{\lambda^{2j}} \\ V(s) &= \frac{\lambda}{\lambda^2+s^2} = \sum_{j=0}^{\infty} \frac{(-1)^j s^{2j+1}}{\lambda^{2j+1}}.\end{aligned}$$

In this case Theorem 2.1.6 can be extended and we have the infinite Taylor expansion

$$\begin{aligned}\psi(s) &= U(s) + iV(s) \\ &= \sum_{j=0}^{\infty} \frac{s^j}{j!}\cdot\frac{i^j j!}{\lambda^j}, \qquad |s| < \lambda.\end{aligned} \qquad \square$$

Theorems 2.1.1 and 2.1.2 assert that uniform convergence is sufficient to allow the order in which certain limits are taken to be altered. However, such interchanges may still be permissible without uniform convergence. In particular we will occasionally change the order in repeated integrals that are not necessarily uniformly convergent, but validity of the operation will usually be apparent from the context.

2.2 POWER SERIES AND PROBABILITY GENERATING FUNCTIONS

Let $A(z)$ denote the power series

$$A(z) = \sum_{n=0}^{\infty} z^n a_n.$$

Later we specialise to the case when $\{a_n\}$ is a probability distribution ($0 \leqslant a_n \leqslant 1$, $\sum_{n=0}^{\infty} a_n = 1$), but it is just as easy to consider the more general situation first.

If the a_n are bounded, say $|a_n| \leqslant B$, then $A(z)$ converges for at least $|z| < 1$ since for these z

$$|A(z)| \leqslant \sum_{n=0}^{\infty} |z^n|\,|a_n| \leqslant \frac{B}{1-|z|}.$$

It may happen that $\sum_n z^n a_n$ converges for some $z > 1$. For example if $a_n = 0$, $n \geqslant m+1$, then $\sum z^n a_n$ is a polynomial of degree m and is convergent for all finite z. The important point is that whether or not the a_n are bounded there is a non-negative $R \leqslant \infty$ such that

$$\sum_n z^n a_n \text{ converges for } |z| < R$$

$$\sum_n z^n a_n \text{ diverges for } |z| > R.$$

The number R is called the *radius of convergence* of the power series $A(z) = \sum z^n a_n$. The interval $(-R, R)$ is called the *interval of convergence*. Many of the results given here hold for complex z but this generality is not necessary for our purposes, and it suffices to take z real in $(-R, R)$ or even in $[0, R)$.

A function $A(z)$ can represent a power series only for values of z within the interval of convergence. For example

$$A(z) = \frac{1}{1-\rho z}$$

with $\rho > 0$ is well defined for all real $z \neq \rho^{-1}$. However, it denotes the geometric series $\sum_{n=0}^{\infty} z^n \rho^n$ only for $|z| < \rho^{-1}$, since if $|z| > \rho^{-1}$ the series diverges.

The fundamental properties of power series can be summarised as follows:

Theorem 2.2.1 Concerning the power series $A(z) = \sum_{n=0}^{\infty} z^n a_n$;

(a) The radius of convergence R is given by

$$R^{-1} = \limsup_{n \to \infty} (a_n)^{1/n}$$

(b) Within the interval of convergence, $A(z)$ is uniformly convergent and has derivatives of all orders which may be obtained by termwise differentiation. Similarly the integral $\int_a^b A(z)\,dz$ is given by termwise integration for any a, b in $(-R, R)$.

(c) (Uniqueness). If $A(z)$ and $B(z) = \sum_{n=0}^{\infty} z^n b_n$ both converge and are equal for all $|z| < R$, then $a_n = b_n, n = 0, 1, 2, \ldots$.

(d) No general statement can be made about the convergence of the series on the boundary of the interval of convergence, i.e. $\sum_n R^n a_n$ may or may not be finite. □

Proofs can be found in most standard books on analysis or infinite series, for example Knopp,* Chapter 5. Part (b) can be deduced from Theorem 2.1.3.

Part (c) implies that there is a 1–1 correspondence between the series $A(z)$ and the sequence $\{a_n\}$ of coefficients. In particular probability distributions with

$$\Pr(X = b+nh) = a_n, \qquad n = 0, 1, 2, \ldots,$$

b and h fixed, are uniquely determined by their generating functions.

* K. Knopp, *Theory and Application of Infinite Series*, Blackie, London, 1928 (trans. of 2nd German edition), first published in 1921 but still one of the best books dealing with series.

From (b) follows the general form of the infinite Taylor expansion of $A(z)$ about the point z_1,

$$A(z) = A(z_1) + \sum_{j=1}^{\infty} \frac{(z-z_1)^j}{j!} A^{(j)}(z_1), \tag{2.2.1}$$

which is valid when $|z-z_1| < R-|z_1|$. Equation (2.1.8) is a particular case.

Part (b) also asserts the uniform convergence for $-R < z < R$ of all series of the form

$$A^{(k)}(z) = \sum_{n=k}^{\infty} (n)_k z^{n-k} a_n, \qquad k = 1, 2, 3, \ldots,$$

where $(n)_k = n(n-1)\ldots(n-k+1)$. If $R \geqslant 1$ and the interchange of limit and summation is valid, then

$$A^{(k)}(1) = \lim_{z\to 1-} A^{(k)}(z) = \sum_{n=k}^{\infty} (n)_k\, a_n. \tag{2.2.2}$$

In the special case when $\{a_n\}$ is the distribution of a random variable X,

$$\Pr(X = n) = a_n, \qquad n = 0, 1, 2, \ldots,$$

we must have $R \geqslant 1$ and the right-hand side of (2.2.2) will be recognised as

$$\mu_{[k]} = E\{X(X-1)\ldots(X-k+1)\}$$

$$= \sum_{n=k}^{\infty} n(n-1)\ldots(n-k+1) \Pr(X=n).$$

Provided the defining series is convergent, $\mu_{[k]}$ is called the *kth factorial moment of* X. Hence if

$$\lim_{z\to 1-} \sum_{n=k}^{\infty} (n)_k z^{n-k} a_n = \sum_{n=k}^{\infty} \lim_{z\to 1-} (n)_k\, z^{n-k} a_n \tag{2.2.3}$$

then

$$A^{(k)}(1) = \mu_{[k]}. \tag{2.2.4}$$

One circumstance in which (2.2.3) is certainly true is when $R > 1$, for then $z = 1$ is an interior point of the convergence interval $(-R, R)$. Furthermore, the conditions of Theorem 2.1.4 are satisfied and from (2.1.8) we have, for values of z near one,

$$A(z) = 1 + \sum_{j=1}^{\infty} \frac{(z-1)^j}{j!} \mu_{[j]}. \tag{2.2.5}$$

All moments are finite* and the probability generating function $A(z)$ generates the factorial moments of X in the sense of (2.2.5).

The question arises as to what happens when $R = 1$—is (2.2.3) still valid? The answer is given by the classical result known as Abel's Theorem which holds for arbitrary real power series. This theorem concerns the behaviour of a power series $A(z)$ at the boundary of its interval of convergence, in particular the limit $\lim_{z \to R-} A(z)$. Without loss of generality assume $R = 1$, for if in fact $R \neq 1$ we consider the series $A_1(z) = \sum_{n=0}^{\infty} z^n R^{-n} a_n$ which does have radius of convergence unity.

Theorem 2.2.2 (Abel) Suppose $A(z) = \sum_{n=0}^{\infty} z^n a_n$ has radius of convergence $R = 1$ and that $\sum_{n=0}^{\infty} a_n$ is convergent. Then

$$\lim_{z \to 1-} A(z) = \sum_{n=0}^{\infty} a_n.$$

If the coefficients a_n are non-negative, the result continues to hold whether or not the sum on the right is convergent.

Proof Suppose firstly that $\sum_{n=0}^{\infty} a_n$ converges to the finite quantity s. Put

$$s_{-1} = 0, \quad s_n = a_0 + a_1 + \ldots + a_n, \qquad n = 0, 1, 2, 3 \ldots.$$

Then for any positive m and $|z| < 1$

$$\begin{aligned} \sum_{n=0}^{m} z^n a_n &= \sum_{n=0}^{m} z^n (s_n - s_{n-1}) \\ &= \sum_{n=0}^{m} z^n s_n - z \sum_{n=0}^{m-1} z^n s_n \\ &= (1-z) \sum_{n=0}^{m} z^n s_n + z^m s_m. \end{aligned}$$

* For a direct proof not using Taylor's Theorem see Exercise 10.

By hypothesis $\lim_{m\to\infty} s_m = s < \infty$, implying (since $|z| < 1$)

$$A(z) = (1-z) \sum_{n=0}^{\infty} z^n s_n. \tag{2.2.6}$$

For arbitrary $\varepsilon > 0$, the convergence of s_n to s means that there exists $N = N(\varepsilon)$ such that

$$|s_n - s| < \tfrac{1}{2}\varepsilon \quad \text{for } n > N.$$

Hence for $0 < z < 1$

$$\begin{aligned} |A(z)-s| &= \left|(1-z) \sum_{n=0}^{\infty} z^n s_n - s\right| = \left|(1-z) \sum_{n=0}^{\infty} z^n (s_n - s)\right| \\ &\leqslant (1-z) \sum_{n=0}^{N} z^n |s_n - s| + (1-z) \sum_{n=N+1}^{\infty} z^n |s_n - s| \\ &\leqslant (1-z) \sum_{n=0}^{N} z^n |s_n - s| + \tfrac{1}{2}\varepsilon. \end{aligned}$$

Let B_N be such that $B_N \geqslant \sum_{n=0}^{N} |s_n - s|$ and let $\delta = \min(1, \varepsilon/2B_N)$. Then for $1-\delta < z < 1$

$$|A(z) - s| \leqslant (1-z)B_N + \tfrac{1}{2}\varepsilon \leqslant \tfrac{1}{2}\varepsilon + \tfrac{1}{2}\varepsilon = \varepsilon.$$

But $\varepsilon > 0$ is arbitrary and so we have proved the first part of Abel's Theorem.

Consider now the case when the a_n are non-negative. This implies that the partial sums s_n are non-decreasing in n, and if the series $\Sigma\, a_n$ is divergent it can only diverge to $+\infty$. The convergent case has already been dealt with so we suppose that

$$\lim_{n\to\infty} s_n = \infty,$$

that is, for arbitrarily large $M > 0$ there exists an integer $N = N(M)$ such that

$$s_n > M \quad \text{for } n > N.$$

We have to show that $\lim_{z\to 1-} A(z) = +\infty$. But by the non-negativity of the a_n, $A(z) \geqslant \sum_{j=0}^{n} z^j a_j$ for every n. Hence, since $A(z)$ is increasing in $z > 0$,

$$\lim_{z \to 1-} A(z) \geqslant \sum_{j=0}^{n} a_j = s_n > M \quad \text{for } n > N,$$

which is what we want to prove. □

An immediate consequence which we will have occasion to use is the

Corollary If the sequence $\{b_n\}$ converges to a limit $b = \lim_{n\to\infty} b_n$ then

$$\lim_{z\to 1-} (1-z) \sum_{n=0}^{\infty} z^n b_n = b. \tag{2.2.7}$$

Proof Interpret b_n as the nth partial sum of the sequence $\{b_j - b_{j-1}\}$ and the corollary follows from (2.2.6) and the subsequent argument. □

The assumption that $\Sigma\, a_n$ is convergent or divergent to $+\infty$ is crucial for Abel's Theorem and the converse proposition

$$\text{'if } \lim_{z\to 1-} \sum_{n=0}^{\infty} z^n a_n = s \qquad \text{then } \lim_{n\to\infty} \sum_{j=0}^{n} a_j = s\text{'}$$

is generally false. A simple counter example is provided by taking

$$a_n = (-1)^n, \qquad n = 0, 1, 2, \ldots.$$

Then

$$A(z) = \sum_{n=0}^{\infty} z^n a_n = (1+z)^{-1}$$

and

$$\lim_{z\to 1-} A(z) = \tfrac{1}{2}.$$

On the other hand

$$s_n = \sum_{j=0}^{n} a_j = \begin{cases} 1 & \text{if } n \text{ is even,} \\ 0 & \text{if } n \text{ is odd,} \end{cases}$$

and $\lim_{n\to\infty} s_n$ does not exist since

$$\liminf_{n\to\infty} s_n = 0 \neq \limsup_{n\to\infty} s_n = 1.$$

Extra conditions on the a_n are required to establish these assertions, and results in this direction are known as *Tauberian Theorems.* Whilst of importance in certain branches of probability theory, they are not necessary for our discussion.

Specializing $A(z) = \sum_{n=0}^{\infty} z^n a_n$ to be the probability generating function of a non-negative random variable X, we have from Abel's Theorem

$$\lim_{z\to 1-} A(z) = 1$$

and

$$\lim_{z\to 1-} A^{(k)}(z) = \sum_{n=k}^{\infty} (n)_k a_n = EX(X-1)\dots(X-k+1)$$

whether or not the right-hand side is convergent. Combining this with (2.2.3) we can finally assert the relationship between the factorial moments $\mu_{[k]}$ of X and the derivatives of the generating function at $z = 1$:

Theorem 2.2.3 If X is a random variable on the non-negative integers with probability generating function $A(z)$ and kth factorial moment $\mu_{[k]}$, then

$$\lim_{z\to 1-} A^{(k)}(z) = \mu_{[k]}, \qquad k = 1,2,\dots,$$

in the sense that both sides of this equation are infinite together or finite and equal. If the radius of convergence R of $A(z)$ is greater than unity then X has finite moments of all orders and (2.2.5) holds, but if $R = 1$, EX^k is finite if and only if $\lim_{z\to 1-} A^{(k)}(z)$ is finite. □

Example 1 Part of Exercise 3 requests that $A(z) = 1-(1-z^2)^{\frac{1}{2}}$ be shown to be a probability generating function. Supposing for the time being that it is the generating function of X, we have from Theorem 2.2.3

$$EX = \lim_{z \to 1-} A'(z) = \lim_{z \to 1-} z(1-z^2)^{-\frac{1}{2}} = \infty,$$

which is a relatively easy way of showing that the series $\sum_{n=1}^{\infty} n \Pr(X = n)$ diverges. □

Abel's Theorem and its consequence Theorem 2.2.3 justify our manipulations with series in § 1.5, and we have achieved a precise statement of the relationship between the moments of a non-negative random variable and the analytic properties of its generating function at $z = 1$. We turn to consider a very different attribute of generating functions which is particularly useful in the discussion of sums of random variables in subsequent chapters. For the moment we treat it as of purely mathematical interest and defer a probabilistic interpretation until Chapter 3.

Suppose $\{a_n\}$ and $\{b_n\}$ are two sequences of numbers indexed by $n = 0, 1, 2, \ldots$, and let c_n be defined for $n = 0, 1, 2, 3, \ldots$, by

$$c_n = \sum_{j=0}^{n} a_{n-j} b_j = \sum_{j=0}^{n} a_j b_{n-j}. \tag{2.2.8}$$

Definition 2.2.1 The sequence $\{c_n\}$ of (2.2.8) is called the *convolution* of the sequences $\{a_n\}$ and $\{b_n\}$, sometimes written $c_n = \{a_n\} * \{b_n\}$. □

Theorem 2.2.4 Suppose $\{c_n\}$ is the convolution of $\{a_n\}$ and $\{b_n\}$ and let $C(z) = \sum_{n=0}^{\infty} z^n c_n$, $B(z) = \sum_{n=0}^{\infty} z^n b_n$, $A(z) = \sum_{n=0}^{\infty} z^n a_n$. Then

$$C(z) = A(z)B(z).$$

Proof. By verification

$$\begin{aligned} A(z)B(z) &= (a_0 + za_1 + z^2 a_2 + \ldots)(b_0 + zb_1 + z^2 b_2 + \ldots) \\ &= a_0 b_0 + z(a_1 b_0 + a_0 b_1) + z^2(a_2 b_0 + a_1 b_1 + a_0 b_2) + \ldots \\ &= \sum_{n=0}^{\infty} z^n \sum_{j=0}^{n} a_{n-j} b_j = C(z). \end{aligned}$$

□

It follows immediately that if $\{a_n\}$, $\{b_n\}$ are probability distributions then their convolution $\{c_n\}$ is also one.

Example 2 $A(z)/(1-z) = (\sum_{n=0}^{\infty} z^n a_n)(\sum_{n=0}^{\infty} z^n) = \sum_{n=0}^{\infty} z^n \sum_{j=0}^{n} a_j.$

Thus the convolution of a known series with the geometric series yields a series whose coefficients are the partial sums. The result appeared in the proof of Abel's theorem as $A(z) = (1-z) \sum_{n=0}^{\infty} z^n s_n$. □

Example 3 We show how Theorem 2.2.4 and (2.2.7) can be used to simplify the calculation of certain limits. Let $\{p_n, n = 1, 2, \ldots\}$ be a probability distribution. Taking $p_0 = 0$ suppose the sequence $\{u_n\}$ is defined recursively by

$$u_0 = 1, \quad u_n = \sum_{j=0}^{n} p_j u_{n-j} \qquad n = 1, 2, \ldots.$$

The problem is to evaluate $\lim_{n\to\infty} u_n$, given that it exists (such sequences appear in § 4.5). Writing $G(z) = \sum_{n=0}^{\infty} z^n p_n$, $U(z) = \sum_{n=0}^{\infty} z^n u_n$, multiplying the equation for u_n by z^n and summing over n, we have by the theorem that $U(z) = 1 + U(z)G(z)$, or

$$U(z) = \frac{1}{1-G(z)}.$$

Applying (2.2.7)

$$\begin{aligned} \lim_{n\to\infty} u_n &= \lim_{z\to 1-} (1-z)U(z) \\ &= \lim_{z\to 1-} \frac{1-z}{1-G(z)} \\ &= \frac{1}{G'(1)} = \frac{1}{\sum_{n=1}^{\infty} n p_n}. \end{aligned}$$

That is, given u_n converges to a limit, its value is the reciprocal of the mean of the distribution $\{p_n\}$. □

Example 4 Let $\Pr(X = n) = a_n, n = 0, 1, 2, \ldots$, and

$$q_n = \Pr(X > n) = \sum_{j=n+1}^{\infty} a_j.$$

Then

$$\begin{aligned} Q(z) &= \sum_{n=0}^{\infty} z^n q_n \\ &= \sum_{j=0}^{\infty} z^n \left(1 - \sum_{j=0}^{n} a_j\right) \\ &= \frac{1}{1-z} - \sum_{n=0}^{\infty} z^n \left(\sum_{j=0}^{n} a_j\right) \\ &= \frac{1-A(z)}{1-z} \end{aligned} \tag{2.2.9}$$

(by Example 2), where $A(z) = \sum z^n a_n$. By Theorem 2.2.3

$$\begin{aligned} EX &= \lim_{z \to 1-} A'(z) = \lim_{z \to 1-} \frac{1-A(z)}{1-z} \\ &= \lim_{z \to 1-} Q(z) = \sum_{n=0}^{\infty} \Pr(X > n). \end{aligned}$$

For non-negative integer-valued random variables we have recovered part of Theorem 1.5.1. □

The natural application of power series in probability theory is to distributions on the non-negative integers. Occasionally one considers series including negative powers of z. For example suppose $0 < \rho < 1$ and that X has distribution

$$\Pr(X = n) = \frac{1-\rho}{1+\rho} \rho^{|n|}, \qquad n = 0, \pm 1, \pm 2, \ldots.$$

The probability generating function

$$\begin{aligned} Ez^X &= \frac{1-\rho}{1+\rho} \sum_{n=-\infty}^{\infty} z^n \rho^{|n|} \\ &= \left[\frac{1-\rho}{1+\rho}\right] \left[\frac{\rho}{z-p} + \frac{1}{1-\rho z}\right] \end{aligned}$$

converges for $-\rho^{-1} < z < -\rho$ and $\rho < z < \rho^{-1}$. Allowing z to be complex, the region of convergence is the annulus $\rho < |z| < \rho^{-1}$, and it is clear that all two-sided lattice distributions have generating functions convergent in an annulus including the unit circle.

For these two-sided distributions it is possibly simpler to consider the moment generating function. In the above example this is

$$Ee^{-sX} = \left[\frac{1-\rho}{1+\rho}\right]\left[\frac{\rho}{e^{-s}-\rho}+\frac{1}{1-\rho e^{-s}}\right]$$

convergent in the interval $\log\rho < s < \log\rho^{-1}$. Furthermore in some cases the annulus in which the generating function is defined shrinks to the unit circle itself. Thus

$$p_n = \frac{12}{\pi^2 n^2}, \qquad n \pm 1, \pm 2, \ldots,$$

has generating function convergent only when $|z| = 1$. Writing $z = e^{is}$, the probability generating function Ez^X becomes the characteristic function Ee^{isX}, defined for all real s. As a result it is generally more convenient to use characteristic functions or moment generating functions when dealing with two sided distributions.

Exercises

Commonly used tests for the convergence of a series $\sum a_n$ are the following:

(a) *Root test*: Let $\alpha = \limsup_{n\to\infty} (|a_n|)^{1/n}$. Then

(i) if $\alpha < 1$, the series is convergent,
(ii) if $\alpha > 1$, the series is divergent,
(iii) if $\alpha = 1$, the test is inconclusive.

α^{-1} will be recognised as the radius of convergence of $\sum_{n=0}^{\infty} z^n a_n$.

(b) *Ratio test*: If for sufficiently large n

(i) $|a_{n+1}/a_n| \leqslant \beta < 1$, the series converges,
(ii) $|a_{n+1}/a_n| \geqslant 1$, for $n > N$, N fixed, the series diverges. The test is inconclusive if

$$\liminf_{n\to\infty} |a_{n+1}/a_n| \leqslant 1 \leqslant \limsup_{n\to\infty} |a_{n+1}/a_n|.$$

(c) *Comparison test*: The series $\sum a_n$, $\sum b_n$ of *positive* terms converge or diverge together if and only if

$$\lim_{n\to\infty} \frac{a_n}{b_n} = c,$$

with c finite and non-zero.

(d) *Integral test*: Suppose the a_n are decreasing in n and let monotone $a(x) = a_n$, $x = n = 0, 1, 2, \ldots$. Then the series $\sum_{n=0}^{\infty} a_n$ and the integral $\int_0^\infty a(x)\,dx$ converge or diverge together.

1. Apply an appropriate test to verify the convergence or divergence of $\sum a_n$ when

$$\text{(a)}\quad a_n = \binom{n}{k}\rho^n, \qquad k \text{ fixed}, \qquad 0 < \rho < 1,$$

$$\text{(b)}\quad a_n = n^{-(1+\delta)}, \qquad n \geqslant 1, \delta \geqslant 0,$$

$$\text{(c)}\quad \sum a_n = \frac{1}{2}+\frac{1}{3}+\frac{1}{2^2}+\frac{1}{3^2}+\frac{1}{2^3}+\frac{1}{3^3}+\ldots,$$

$$\text{(d)}\quad a_n = \frac{1}{n(\log n)^\delta}, \qquad \delta > 0,$$

$$\text{(e)}\quad a_{2n} = \binom{\frac{1}{2}}{n}(-1)^{n-1}, \qquad a_{2n+1} = 0.$$

2. If $a_n > 0$ show that

$$\begin{aligned}\liminf_{n\to\infty} a_{n+1}/a_n &\leqslant \liminf_{n\to\infty} (a_n)^{1/n} \\ &\leqslant \limsup_{n\to\infty} (a_n)^{1/n} \\ &\leqslant \limsup_{n\to\infty} a_{n+1}/a_n.\end{aligned}$$

Apply these inequalities to $a_n = \cos n$, $a_n = n\sin(\frac{1}{3}n\pi)$.

3. Which of the following power series are probability generating functions:

$$\text{(a)}\quad 1-\{1-4p(1-p)z^2\}^{\frac{1}{2}}, \qquad 0 < p < 1,$$

$$\text{(b)}\quad \left(\frac{1-\rho}{1-\rho z}\right)^\delta, \qquad \delta > 0, 0 < \rho < 1,$$

$$\text{(c)}\quad (1-\rho+\rho z)^\delta, \qquad \delta > 0, 0 < \rho < 1,$$

(d) $e^{\lambda(G(z)-1)}$, $\quad \lambda > 0$, $G(z)$ a probability generating function.

4. Suppose X is a random variable with probability generating function $G(z)$. Write down the probability generating functions of $3X$, $X+1$, $aX+b$, a, b constants.

5. Verify that the following define probability distributions and find the corresponding generating functions.

(a) $p_j = \binom{-N}{j} p^N (p-1)^j, \quad N \text{ fixed}, j = 0, 1, 2, \ldots, \quad 0 < p < 1,$

(b) $p_j = \sum_{k=0}^{j} (1-\rho_1)(1-\rho_2)\rho_1^k \rho_2^{j-k}, \quad 0 < \rho_1, \rho_2 < 1,$
$j = 0, 1, 2, \ldots.$

(c) $\Pr(X=j)$, where X is the number shown when a fair die is rolled.

6. Find the generating functions $\Pr(X = n+1)$, $\Pr(X < n)$, where X is a random variable on the non-negative integers.

7. The generating function $Q(z)$ of the tail sequence $\{\Pr(X > n)\}$ is given by (2.2.9). If $q_n = \Pr(X > n)$ let $r_n = \sum_{j=n+1}^{\infty} q_j$. Find the generating function $R(z)$ of $\{r_n\}$ and relate $R(1)$ to EX^2.

8. Extend the convolution theorem as follows. Suppose $\{a_n^{(1)}\}$, $\{a_n^{(2)}\}, \ldots, \{a_n^{(k)}\}$, have generating functions $A_i(z)$, $i = 1, 2, \ldots, k$ respectively. Show by induction that the generating function of $\{a_n^{(1)}\} * \{a_n^{(2)}\} * \ldots * \{a_n^{(k)}\}$ is $\prod_{i=1}^{k} A_i(z)$. If each $\{a_n^{(j)}\}$ is a probability distribution conclude that their k-fold convolution is also one.

9. (Continuation.) In particular, if all k sequences are equal the generating function of their k-fold convolution is the kth power of their common generating function. Verify this with $k = 3$ and $a_n = (1-\rho)\rho^n$, $n = 0, 1, 2, \ldots$. Compare the mean and variance of the convolution with the original first two moments.

10. Suppose the probability generating function

$$G(z) = \sum_{n=0}^{\infty} z^n \Pr(X = n)$$

has radius of convergence $R > 1$. Show that there is a positive constant c and an α, $0 < \alpha < 1$, such that

$$\Pr(X > n) \leqslant c\alpha^n, \qquad n = 1, 2, 3, \ldots.$$

Conclude that for such random variables $\sum_{n=1}^{\infty} n^k \Pr(X > n)$ is convergent for every $k = 1, 2, 3, \ldots$, and hence, by Theorem 1.5.1, X has moments of all orders.

11. Prove that if $\lim_{n\to\infty} b_n = b$ then $\lim_{n\to\infty} n^{-1} \sum_{j=1}^{n} b_j = b$. A generalisation for non negative sequences $\{a_n\}$, $\{b_n\}$ is

$$\lim_{n\to\infty} \left(\sum_{j=0}^{n} a_j b_{n-j} \Big/ \sum_{j=0}^{n} a_j\right) = b$$

where $b = \lim_{n\to\infty} b_n$ provided $a_n\{\sum_{j=0}^{n} a_j\}^{-1} \to 0$. This second result is useful in obtaining certain properties of Markov chains (e.g. Exercise 5.5.6).

2.3 INTEGRAL TRANSFORMS

If X is a random variable on $(-\infty, \infty)$ with density $f(x)$, its moment generating function and characteristic function

$$M(s) = Ee^{-sX} = \int_{-\infty}^{\infty} e^{-sx} f(x)\, dx$$

$$\psi(s) = Ee^{isX} = \int_{-\infty}^{\infty} e^{isx} f(x)\, dx$$

are examples of integral transforms. $M(s)$ is sometimes called the *bilateral Laplace transform* of $f(x)$ in distinction to the half-line integral on $[0, \infty)$ which is usually referred to simply as the Laplace transform. The characteristic function $\psi(s)$ is also called the *Fourier transform* of $f(x)$.

As with probability generating functions (Theorem 2.2.1c), there is a 1–1 correspondence between integral transforms and random variables in the sense that if two transforms are equal throughout their region of convergence then the corresponding distributions are the same. This is a result of fundamental importance. The proof, however, requires analytic techniques beyond our scope, and we merely state the general result.

Theorem 2.3.1 The moment generating function $M(s) = Ee^{-sX}$ or the characteristic function $\psi(s) = Ee^{isX}$ uniquely determine the distribution function of X at all its points of continuity. □

There is a substantial literature on the subject of integral transforms but, as with power series, only certain facets of the theory are relevant to our interests, namely

(a) their use to generate moments,
(b) the transforms of convolutions.

We commence with the latter.

Equation (2.2.8) defined $c_n = \sum_{j=0}^{n} a_{n-j}b_j, n = 0, 1, 2, \ldots$, to be the convolution of the sequences $\{a_n\}$ and $\{b_n\}$. Similarly

Definition 2.3.1 The function

$$c(x) = \int_{-\infty}^{\infty} a(x-y)b(y)\,dy = \int_{-\infty}^{\infty} a(y)b(x-y)\,dy, \qquad -\infty < x < \infty,$$

is called the *convolution* of the functions $a(x)$ and $b(x)$. □

Theorem 2.3.2 Let $\psi_a(s)$ and $\psi_b(s)$ be the Fourier transforms of the absolutely integrable functions $a(x)$ and $b(x)$

$$\psi_a(s) = \int_{-\infty}^{\infty} e^{isx}a(x)\,dx, \qquad \psi_b(s) = \int_{-\infty}^{\infty} e^{isx}b(x)\,dx.$$

If $c(x)$ is the convolution of $a(x)$ and $b(x)$ its Fourier transform, $\psi_c(s)$ is given by

$$\psi_c(s) = \psi_a(s)\psi_b(s).$$

Proof.

$$\begin{aligned}\psi_c(s) &= \int_{-\infty}^{\infty} e^{isx}c(x)\,dx \\ &= \int_{-\infty}^{\infty}\int_{-\infty}^{\infty} e^{isx}a(x-y)b(y)\,dydx\end{aligned}$$

$$= \int_{-\infty}^{\infty}\int_{-\infty}^{\infty} e^{is(u+v)}a(u)b(y)\,dydu$$

$$= \int_{-\infty}^{\infty} e^{isu}a(u)\,du \int_{-\infty}^{\infty} e^{isv}b(y)\,dy$$

$$= \psi_a(s)\psi_b(s). \qquad \square$$

Precisely the same argument applies to Laplace transforms.

Corollary If $M_a(s)$, $M_b(s)$, and $M_c(s)$ are the bilateral Laplace transforms of the functions $a(x)$, $b(x)$ and $c(x)$ of the theorem, then

$$M_c(s) = M_a(s)M_b(s)$$

for those values of s for which both integrals on the right are convergent. □

Example 1 Let $X \frown N(\mu, \sigma^2)$. That is (page 32), X has density

$$f(x) = \frac{1}{\sigma(2\pi)^{\frac{1}{2}}}\exp\left[\frac{-(x-\mu)^2}{2\sigma^2}\right], \qquad -\infty < x < \infty.$$

The moment generating function is (cf. Exercise 1.5.9a)

$$M(s) = \frac{1}{\sigma(2\pi)^{\frac{1}{2}}}\int_{-\infty}^{\infty} \exp\left\{-sx-\frac{(x-\mu)^2}{2\sigma^2}\right\}dx$$

$$= \frac{1}{\sigma(2\pi)^{\frac{1}{2}}}\int_{-\infty}^{\infty} \exp\left[-\frac{1}{2\sigma^2}\{x^2-2(\mu-s\sigma^2)x+\mu^2\}\right]dx$$

$$= \frac{1}{\sigma(2\pi)^{\frac{1}{2}}}\int_{-\infty}^{\infty} \exp\left[-\frac{1}{2\sigma^2}\{(x-\mu+s\sigma^2)^2-(\mu-s\sigma^2)^2+\mu^2\}\right]dx$$

$$= \exp(-s\mu+\tfrac{1}{2}s^2\sigma^2)\,.\,\frac{1}{\sigma(2\pi)^{\frac{1}{2}}}\int_{-\infty}^{\infty} \exp\left[-\frac{\{x-(\mu-s\sigma^2)\}^2}{2\sigma^2}\right]dx$$

$$= \exp(-s\mu+\tfrac{1}{2}s^2\sigma^2),$$

since the second term is the integral of a $N(\mu-s\sigma^2, \sigma^2)$ density and is therefore equal to unity. Replacing s by $-is$ yields the characteristic function

$$\psi(s) = \exp(is\mu-\tfrac{1}{2}s^2\sigma^2).$$

For this random variable both $M(s)$ and $\psi(s)$ converge for all real s. The convolution of a $N(\mu_1, \sigma_1^2)$ and a $N(\mu_2, \sigma_2^2)$ density has characteristic function $\exp\{is(\mu_1 + \mu_2) - \frac{1}{2}s^2(\sigma_1^2 + \sigma_2^2)\}$ and is therefore a $N(\mu_1 + \mu_2, \sigma_1^2 + \sigma_2^2)$ density. □

Example 2 Given the n functions $a_j(x), j = 1,2,..,n$, their n-fold convolution $c_n(x)$ is defined recursively by

$$c_2(x) = \int_{-\infty}^{\infty} a_1(x-y)a_2(y)\, dy$$

$$c_3(x) = \int_{-\infty}^{\infty} c_2(x-y)a_3(y)\, dy$$

$$\vdots$$

$$c_n(x) = \int_{-\infty}^{\infty} c_{n-1}(x-y)a_n(y)\, dy.$$

If $\gamma_n(s) = \int_{-\infty}^{\infty} e^{isx}c_n(x)\, dx$ and $\psi_j(s) = \int_{-\infty}^{\infty} e^{isx}a_j(x)\, dx, j = 1,2,\ldots,n$, repeated application of Theorem 2.3.2 shows that

$$\gamma_n(s) = \prod_{j=1}^{n} \psi_j(s).$$

An important special case is when the $a_j(x)$ are all equal to, say, $a(x)$, $c_n(x)$ becomes the *n-fold iterated convolution* of $a(x)$ and

$$\gamma_n(s) = \left\{\int_{-\infty}^{\infty} e^{isx}a(x)\, dx\right\}^n,$$

a result used extensively in Chapters 3 and 4. □

Example 3 In Example 2 take $a(x)$ as the negative exponential density function

$$a(x) = \lambda e^{-\lambda x}, \qquad 0 < x.$$

From Example 1.5.7 the moment generating function is

$$M(s) = \int_0^{\infty} e^{-sx}\lambda e^{-\lambda x}\, dx = \lambda/(s+\lambda), \qquad s > -\lambda.$$

The n-fold iterated convolution has transform $\{\lambda/(s+\lambda)\}^n$ and this turns out to be (Exercise 2) the transform of the density function

$$c_n(x) = \frac{e^{-\lambda x}\lambda^n x^{n-1}}{(n-1)!}, \qquad 0 < x. \qquad \square$$

Example 4 It is possible to combine Definitions 2.2.1 and 2.3.1 to define the convolution of a continuous integrable function $a(x)$ and a sequence $\{b_n\}$ by

$$c(x) = \sum_y a(x-y)b_y,$$

the summation being over all y for which $b_y \neq 0$ and $a(x-y) \neq 0$. The argument of Theorem 2.3.2 can be modified to show that

$$\int_{-\infty}^{\infty} e^{isx}c(x)\,dx = \left\{\int_{-\infty}^{\infty} e^{isx}a(x)\,dx\right\}\left\{\sum_x e^{isx}b_x\right\}.$$

An important property of such convolutions is that the resulting function $c(x)$ is always continuous provided $\sum_y b_y$ is absolutely convergent. For the continuity of $a(x)$ implies the uniform convergence of $\sum_y a(x-y)\,b_y$, and hence

$$\lim_{\delta\to 0} |c(x+\delta)-c(x)| \leqslant \sum_y |b_y| \lim_{\delta\to 0} |a(x-y+\delta)-a(x-y)| = 0.$$

We have here a case of the 'smoothing' effect of convolution. $\square$

Example 5 It is apparent that the convolution of two density functions is itself a density function (one has only to check that it integrates to unity and is non-negative). As an application we show that given any density function $f(x)$ we can always obtain a density symmetric about the origin by convolving $f(x)$ with $f(-x)$. Thus

$$c(x) = \int_{-\infty}^{\infty} f(x-y)f(-y)\,dy = \int_{-\infty}^{\infty} f(x+u)f(u)\,du,$$

and by a simple change of variable, $c(x) = c(-x)$. If $\psi(s) = Ee^{isX}$ is the characteristic function of $f(x)$, the characteristic function of $c(x)$ is

$$\int_{-\infty}^{\infty} e^{isx}c(x)\,dx = (Ee^{isX})(Ee^{-isX})$$

$$= |\psi(s)|^2,$$

G

where $k = M(s_0) + M(s_1)$ and $s_2 = min(s_0, s_1)$. Thus the probability origin have real characteristic functions, as indeed can been seen directly from the definition. □

The probabilistic interpretation of convolution is deferred until the next chapter, and we proceed to discuss some aspects of the relationship between transforms of density functions and moments. Consider first moment generating functions.

Theorem 2.3.3 If the integral defining the moment generating function $M(s) = Ee^{-sX}$ converges in an interval $(-\gamma,\gamma)$ including the origin, then X has moments of all orders.

Proof. Exercise 2.2.10 treated the special case of a discrete non-negative X. Supposing that X has density $f(x)$, we are given that the integral

$$M(s) = \int_{-\infty}^{\infty} e^{-sx}f(x)\,dx = \int_{0}^{\infty} e^{-sx}f(x)\,dx + \int_{-\infty}^{0} e^{-sx}f(x)\,dx$$

converges for $-\gamma < s < \gamma$, implying that

$$\int_{0}^{\infty} e^{-sx}f(x)\,dx < \infty \quad \text{for } -\gamma < s,$$

and

$$\int_{-\infty}^{0} e^{-sx}f(x)\,dx < \infty \quad \text{for } \gamma > s.$$

Choose a positive number s_0 such that $-\gamma < -s_0 < 0$. Then

$$\begin{aligned} M(-s_0) &\geqslant \int_{0}^{\infty} e^{s_0 y}f(y)\,dy \\ &\geqslant \int_{x}^{\infty} e^{s_0 y}f(x)\,dy, \quad \text{for every } x \geqslant 0, \\ &\geqslant e^{s_0 x}\int_{x}^{\infty} f(y)\,dy, \quad \text{since } e^{s_0 y} \text{ is } \uparrow \text{ in } y, \\ &= e^{s_0 x}\Pr(X > x). \end{aligned}$$

Therefore

$$\Pr(X > x) \leqslant M(-s_0)e^{-s_0 x} \tag{2.3.1}$$

and for every $r > 0$

$$\int_0^\infty x^{r-1} \Pr(X > x)\, dx \leqslant M(-s_0) \int_0^\infty e^{-s_0 x} x^{r-1}\, dx < \infty.$$

Similarly choose s_1 in $(0, \gamma)$. Then the same argument yields

$$M(s_1) \geqslant \int_{-\infty}^0 e^{-s_1 y} f(y)\, dy \geqslant e^{-s_1 x} \Pr(X < x),$$

and for $x < 0$

$$\Pr(X < x) \leqslant e^{s_1 x} M(s_1), \tag{2.3.2}$$

$$\int_{-\infty}^0 |x|^{r-1} \Pr(X < x)\, dx \leqslant M(s_1) \int_{-\infty}^0 e^{s_1 x} |x|^{r-1}\, dx$$

$$= M(s_1) \int_0^\infty e^{-s_1 y} y^{r-1}\, dy < \infty$$

Consequently for every $r > 0$

$$\int_{-\infty}^\infty |x|^{r-1} \Pr(|X| > x)\, dx$$

$$= \int_0^\infty x^{r-1} \Pr(X > x)\, dx + \int_{-\infty}^0 |x|^{r-1} \Pr(X < x)\, dx$$

is convergent, which by Theorem 1.5.1. implies that EX^r is well defined. □

The argument to obtain (2.3.1) and (2.3.2) illustrates an important technique which was used previously in establishing Chebyshev's Inequality (1.5.4). Given a moment generating function, we have in fact extended the classical result to obtain exponentially diminishing bounds on the tail $\Pr(|X| > x)$ of a distribution. Adding (2.3.1) and (2) and recalling that $s_0, s_1 > 0$ we find for $x > 0$

$$\begin{aligned} \Pr(|X| > x) &= \Pr(X > x) + \Pr(X < -x) \\ &\leqslant M(s_0)e^{-s_0 x} + M(s_1)e^{-s_1 x} \\ &\leqslant ke^{-s_2 x}, \end{aligned} \tag{2.3.3}$$

where $k = M(s_0) + M(s_1)$ and $s_2 = min(s_0, s_1)$. Thus the probability of obtaining very large or very small values of X is small in this sense. The result is of sufficient interest to state separately and in greater generality.

Theorem 2.3.4 If X has a moment generating function convergent in an interval including the origin, then the exponential inequalities (2.3.1)–(2.3.3) hold. Conversely, if (2.3.3) holds for x sufficiently large then X has a moment generating function. □

We have seen in § 2.1 that for $s > -\gamma$

$$\frac{d^k}{ds^k}\int_0^\infty e^{-sx}f(x)\,dx = \int_0^\infty (-1)^k x^k e^{-sx} f(x)\,dx$$

i.e. it is possible to differentiate under the integral sign for s within the region of convergence. In the same way

$$\frac{d^k}{ds^k}\int_{-\infty}^0 e^{-sx}f(x)\,dx = \int_{-\infty}^0 (-1)^k x^k e^{-sx} f(x)\,dx$$

if $s < \gamma$, since the region of convergence of the integral on the left is $s < \gamma$. Putting these together we have

$$\frac{d^k}{ds^k}\int_{-\infty}^\infty e^{-sx}f(x)\,dx = \int_{-\infty}^\infty (-1)^k x^k e^{-sx} f(x)\,dx, \qquad -\gamma < s < \gamma. \tag{2.3.4}$$

In particular (2.3.4) holds at $s = 0$, and we identify

$$M^{(k)}(0) = (-1)^k \int_0^\infty x^k f(x)\,dx = (-1)^k EX^k. \tag{2.3.5}$$

The Taylor expansion (2.1.10) of Theorem 2.1.5 can then be generalised and rewritten as

Theorem 2.3.5 If the moment generating function $M(s) = Ee^{-sX}$ converges for $-\gamma < s < \gamma$ the moments $EX^j, j = 1, 2, 3, \ldots,$ are given by (2.3.5) and

$$M(s) = 1 + \sum_{j=1}^{\infty} \frac{(-s)^j}{j!} EX^j, \qquad -\gamma < s < \gamma. \quad \square$$

Example 6 If $X \frown N(0, \sigma^2)$ then from Example 1

$$Ee^{-sX} = e^{\frac{1}{2}s^2\sigma^2} = 1 + \sum_{j=1}^{\infty} \frac{s^{2j}}{(2j)!} \frac{(2j)!\sigma^{2j}}{j!2^j}, \quad -\infty < s < \infty.$$

Moments of odd order vanish (the density is symmetric about the origin) and

$$EX^{2j} = \frac{(2j)!\sigma^{2j}}{j!2^j}, \qquad j = 1, 2, 3, \ldots. \qquad \square$$

For random variables on a half line, say $[0, \infty)$, the density $f(x)$ is zero for $x < 0$ and the moment generating function reduces to

$$M(s) = \int_0^{\infty} e^{-sx} f(x)\, dx.$$

The range of integration is taken as $[0, \infty)$ and we should perhaps have written the integral more precisely with $\int_{0-}^{\infty}$. Such $M(s)$ converge for all $s \geqslant 0$ but may not be convergent for any negative of s.

An example is

$$M(s) = \int_0^{\infty} e^{-sx} \frac{(m+1)}{(1+x)^{m+2}}\, dx,$$

defined only for $s \geqslant 0$. This function is differentiable infinitely often if $s > 0$ but has only m derivatives at $s = 0$ (c.f. Exercise 1.5.8 and Example 2.1.1). We want to identify

$$\lim_{s \to 0+} M^{(r)}(s) = (-1)^r EX^r, \qquad r = 1, 2, \ldots, m,$$

and to do this we need an Abelian result corresponding to Theorem 2.2.2 for power series.

Theorem 2.3.6 (Abel) Let $g(x)$ be zero if $x < 0$ and suppose $M_g(s) = \int_0^{\infty} e^{-sx} g(x)\, dx$ converges if $s > 0$. If $G(x) = \int_0^x g(y)\, dy$ and

$$\lim_{x \to \infty} G(x) = G = \int_0^{\infty} g(y)\, dy$$

then

$$\lim_{s \to 0+} M_g(s) = G.$$

Proof. An integration by parts shows that

$$M_g(s) = s\int_0^\infty e^{-sx}G(x)\,dx. \tag{2.3.6}$$

Recognising that $\int_0^\infty e^{-sx}\,dx = s^{-1}$, we have

$$M_g(s) - G = s\int_0^\infty e^{-sx}(G(x)-G)\,dx$$

and hence for any $N > 0$

$$|M_g(s)-G| \leqslant s\int_0^N e^{-sx}|G(x)-G|\,dx + s\int_N^\infty e^{-sx}|G(x)-G|\,dx.$$

Since $\lim_{x\to\infty} G(x) = G$ it is possible to choose $N = N(\varepsilon)$ so large that for arbitrary $\varepsilon > 0$

$$|G(x)-G| < \varepsilon \quad \text{for } x > N.$$

Then

$$|M_g(s)-G| \leqslant s\int_0^N e^{-sx}|G(x)-G|\,dx + \varepsilon e^{-sN},$$

and letting s approach zero we find

$$\lim_{s\to 0+} |M_g(s)-G| \leqslant \varepsilon.$$

But ε can be made arbitrarily small, and the proof is complete. □

Corollary If $a(x)$ is defined on the right half-line, is differentiable, and $\lim_{x\to\infty} a(x) = a$, then

$$\lim_{s\to 0+} s\int_0^\infty e^{-sx}a(x)\,dx = a.$$

Proof. Substitute $G(x) = a(x)$ in (2.3.6) and use the fact that

$$\lim_{x\to\infty} a(x) = \lim_{x\to\infty}\int_0^x a'(y)\,dy. \quad \square$$

Clearly if $g(x)$ is non-negative (and hence $G(x)$ non-decreasing) we have the analogue to the second part of Theorem 2.2.2, namely

$$\lim_{s\to 0+} M_g(s) = \infty \quad \text{if} \quad \lim_{x\to\infty} G(x) = \infty.$$

Taking $g(x) = x^j f(x)$, the version of Abel's Theorem for Laplace transforms therefore yields

Theorem 2.3.7 If X is a non-negative random variable with moment generating function $M(s) = Ee^{-sx}$ then

$$(-1)^j \frac{d^j}{ds^j} M(s)\Big|_{s=0} = EX^j, \qquad j = 1, 2, 3, \ldots,$$

in the sense that both sides are infinite together or finite and equal. □

These last two theorems also apply to random variables whose densities can be non zero only on $(-\infty, 0]$, the only changes of proof required being the replacement of $\int_0^\infty$ by $\int_{-\infty}^0$ and $\lim\limits_{s \to 0+}$ by $\lim\limits_{s \to 0-}$.

Returning to random variables on $(-\infty, \infty)$ whose densities do not necessarily vanish on a half line, recall that the moment generating function

$$M(s) = \int_0^\infty e^{-sx} f(x)\, dx + \int_{-\infty}^0 e^{-sx} f(x)\, dx$$

is defined for $-\gamma < s < \gamma$ if the first integral on the right converges for at least $s > -\gamma$, and the second for at least $s < \gamma$. If in fact these integrals converge only for $s \geqslant 0$ and $s \leqslant 0$ respectively, then a moment generating function does not exist since for real argument $M(s)$ converges only at $s = 0$. Such circumstances oblige one to replace $-s$ by a complex variable and to use the characteristic function

$$\psi(s) = \int_{-\infty}^\infty e^{isx} f(x)\, dx.$$

The additional generality gained by the fact that a characteristic function exists for all random variables carries with it the penalty of increased technical complexity. In particular the relationship between derivatives $\psi^{(j)}(0)$ and moments EX^j is less clear cut than in the case of moment generating functions. It turns out to be true that if EX^j exists then $\psi(s)$ is j times differentiable at the origin, but the converse is not necessarily the case unless j is even. We do not discuss the question in detail but give only a partial argument.

The first derivative

$$\psi'(s) = \frac{d}{ds}\int_{-\infty}^{\infty} \cos sx\, f(x)\, dx + i\frac{d}{ds}\int_{-\infty}^{\infty} \sin sx\, f(x)\, dx$$

becomes

$$\psi'(s) = -\int_{-\infty}^{\infty} x \sin sx\, f(x)\, dx + i\int_{-\infty}^{\infty} x\cos sx\, f(x)\, dx$$

if it is permissible to differentiate under the integral sign, and at $s = 0$ this reduces to

$$\psi'(0) = i\int_{-\infty}^{\infty} x\, f(x)\, dx = iEX. \tag{2.3.7}$$

A sufficient condition that ensures the interchange of operations is that

$$E|X| = \int_{-\infty}^{\infty} |x|\, f(x)\, dx < \infty,$$

namely that EX exists. To see this note that

$$\begin{aligned}\left|\int_{-\infty}^{\infty} (1-\cos sx)x\, f(x)\, dx\right| & \\ \leqslant \int_{-b}^{b} |1-\cos sx|\,|x|\, f(x)\, dx &+ \int_{|x|\geqslant b} |x|\, f(x)\, dx \\ \to \int_{|x|\geqslant b} |x|\, f(x)\, dx \quad & \text{as } s \to 0.\end{aligned}$$

But the absolute convergence of $xf(x)$ implies that b can be chosen to make the last integral arbitrarily small.

The argument generalises and without further detail we state

Theorem 2.3.8 If $\psi(s)$ is the characteristic function of X whose jth moment EX^j exists then

$$\psi^{(j)}(0) = i^j EX^j. \qquad \square$$

Combining this result with Theorem 2.1.6 we finally have

Theorem 2.3.9 If X has an rth moment then $\psi(s) = Ee^{isX}$ can be expanded in the Taylor series

$$\psi(s) = 1 + \sum_{j=1}^{r} \frac{(is)^j}{j!} EX^j + o(s^r), \qquad s \to 0. \qquad \square$$

One point concerning generating functions remains to be discussed.* The question is, given a transform how does one retrieve the corresponding distribution function? Expressions which give $F(x)$ in terms of $\psi(s)$ are called *inversion formulae.* We do not have occasion to use these formulae, and rely when necessary on the table of generating functions given at the end of this Chapter. However, partly because of its brevity but also by way of illustration, we derive an inversion formula for lattice distributions.

Let X be lattice with span h. The only values X can take with non-zero probability are of the form

$$x = b + kh, \qquad k = 0, \pm 1, \pm 2, \ldots,$$

with b and h given. For convenience take $b = 0$ (or otherwise consider $Y = X - b$ whose characteristic function is

$$Ee^{isY} = e^{-isb} Ee^{isX}).$$

If

$$p_k = \Pr(X = kh), \qquad k = 0, \pm 1, \pm 2, \ldots,$$

the characteristic function of X is

$$\begin{aligned} \psi(s) &= \sum_{k=-\infty}^{\infty} e^{ihsk} p_k \\ &= \sum_{k=-\infty}^{\infty} p_k \cos hsk + i \sum_{k=-\infty}^{\infty} p_k \sin hsk. \qquad (2.3.8) \end{aligned}$$

The trigonometric functions ensure that $\psi(s)$ is periodic with period $2\pi/h$, in particular, $\psi(s) = 1$ whenever

$$s = 2\pi n/h, \qquad n = 0, \pm 1, \pm 2, \ldots.$$

It suffices then to consider $\psi(s)$ for s in an interval of length $2\pi/h$, say $-(\pi/h) \leqslant s \leqslant (\pi/h)$.

* We take up briefly an important but special topic the results of which will be used only in the latter part of § 4.3.

For any integer j

$$\int_{-\pi/h}^{\pi/h} e^{ihsj}\,ds = \frac{2\sin \pi j}{hj} = \begin{cases} 0 & \text{if } j \neq 0, \\ \dfrac{2\pi}{h} & \text{if } j = 0, \end{cases}$$

and if we integrate the identity

$$e^{-ihsn}\psi(s) = p_n + \sum_{k \neq n} p^k e^{ihs(k-n)}$$

with respect to s over $(-\pi/h, \pi/h)$ we find

$$\int_{-\pi/h}^{\pi/h} e^{-ihsn}\psi(s)\,ds = 2\pi p_n/h.$$

Hence the inversion formula

Theorem 2.3.10 If X is lattice with span h and characteristic function $\psi(s)$ then

$$\Pr(X = nh) = \frac{h}{2\pi}\int_{-\pi/h}^{\pi/h} e^{-ihsn}\psi(s)\,ds, \qquad n = 0, \pm 1, \pm 2, \ldots \quad \square$$

Exercises

1. Find the moment generating functions (and characteristic functions) of the densities

(a) $2^{-1}e^{-|x|}, \quad -\infty < x < \infty,$
(b) $(2c)^{-1}, \quad -c < x < c.$

2. (Continuation of Example 3.) Show by integration and induction that the n-fold iterated convolution of the negative exponential density $a(x) = \lambda e^{-\lambda x}, 0 < x$, is the so called *gamma* density (c.f. 2.5.5)

$$c_n(x) = \frac{e^{-\lambda x}\lambda^n x^{n-1}}{(n-1)!}, \qquad 0 < x,$$

whose moment generating function is $\{\lambda/(\lambda+s)\}^n$. Find its mean and variance.

3. Recover (2.3.6) by taking $b(y) = 0$ if $y < 0$, $a(x) = 1$ if $0 < x$, $a(x) = 0$ otherwise, in the Laplace transform of the convolution $c(x) = \int_{-\infty}^{\infty} a(x-y)b(y)\,dy$.

4. Find the characteristic function and density function of the convolution of the densities of X and $Y = -X$ when

(a) $X \frown N(\mu, \sigma^2)$,
(b) X is uniform on $(0, c)$ i.e. $f(x) = c^{-1}, 0 < x < c$.

Care must be taken with the range of integration in (b).

5. If $F(x)$ is the distribution function of the positive random variable X with finite mean μ and moment generating function $M(s) = Ee^{-sX}$, show that $f_{(1)}(x) = \mu^{-1}(1-F(x)), 0 < x$, is a density function and that its moment generating function is $(\mu s)^{-1}(1-M(s))$.

6. (Continuation.) If $F_{(1)}(x) = \int_0^x f_{(1)}(y)\,dy$ and $\mu_2 = EX^2$, find the constant c which ensures that $f_{(2)}(x) = c\{1-F_{(1)}(x)\}$ is also a density function. Write down the moment generating function of $f_{(2)}(x)$ and find its mean and variance.

7. (Continuation.) Show that the negative exponential is the only density satisfying $f(x) = f_{(2)}(x)$, where $f(x) = F'(x)$.

8. If $f(x)$ is a density on $(0, \infty)$ use Abel's Theorem to find $\lim_{x\to\infty} g(x)$ when $g(x) = f(x) + \int_0^x f(x-y)g(y)\,dy$, given that the limit exists.

9. For what values of c_1 and c_2 is $2^{-1}\{(1+c_1 s)^{-1}+(1+c_2 s)^{-1}\}$ the moment generating function of a positive random variable?

10. If $\{F_n(x)\}$ is a sequence of distribution functions with characteristic functions $\{\psi_n(s)\}$ show that if $a_n \geqslant 0$ and $\sum_{n=0}^{\infty} a_n = 1$ then

$$F(x) = \sum_{n=0}^{\infty} a_n F_n(x)$$

is also a distribution function whose characteristic function is $\psi(s) = \sum_{n=0}^{\infty} a_n \psi_n(s)$.

11. (Continuation.) If $F_n'(x) = f_n(x)$ and $f_n(x)$ is the n-fold iterated convolution of $f_1(x)$, find $\psi(s)$ when $a_n = e^{-\lambda}\lambda^n/(n!)$.

12. It is apparent that the theory for $\sum_{n=0}^{\infty} e^{-sn}p_n$ has much in common with that for $\int_0^\infty e^{-sx}f(x)\,dx$. However, arguments concerning integrals are usually more complicated. One reason is that a function may be integrable over the positive half-line yet not tend to zero as $x \to \infty$, whereas the nth term of a convergent series necessarily approaches 0. Construct an example of such non-negative integrable $g(x)$ using a 'saw tooth' function with

$$\limsup_{x\to\infty} g(x) = 1, \qquad \liminf_{x\to\infty} g(x) = 0,$$

and which vanishes everywhere except in suitable intervals about the integers $n = 1, 2, 3, \ldots$.

13. Marcinkiewicz's theorem states that if $e^{Q(s)}$ is a characteristic function with $Q(s)$ a polynomial then necessarily the degree of $Q(s)$ cannot exceed two. Show by a direct argument that $\exp(bs^3)$ cannot be a characteristic function.

14. The logarithm $K(s) = \log(Ee^{isX})$ of the characteristic function Ee^{isX} is called the *cumulant generating function of* X. If a power series development

$$K(s) = \sum_{j=1}^{\infty} \frac{(is)^j}{j!} K_j$$

is possible, the coefficients K_j are called *cumulants* or *semi invariants.* Show that $K_1 = EX$ and $K_2 = \operatorname{Var} X$.

15. Show that if $f(x)$ and $\psi(s)$ are respectively the density and characteristic function of a $N(0, 1)$ variable then

$$f(x) = \frac{1}{2\pi}\int_{-\infty}^{\infty} e^{-isx}\psi(s)\,ds.$$

(This inversion formula in fact holds for all integrable characteristic functions).

16. Obtain the Taylor expansion of Theorem 2.3.9 for the characteristic function of the two-fold convolution of the density $f(x) = 3x^{-4}, 1 \leqslant x$.

17. Let $\psi(s)$ be the characteristic function of a lattice variable with span h. Show that for given $\delta > 0$ it is possible to find $\gamma > 0$ such that $|\psi(s)| \leqslant e^{-\gamma}$ if $\delta \leqslant |s| \leqslant \pi/h$.

18. Use the inversion formula of Theorem 2.3.10 to show that if $\psi(s) = \cos s$ then $\Pr(X = \pm 1) = \frac{1}{2}$, $\Pr(X = j) = 0$ otherwise.

19. A classical problem is whether or not moments uniquely determine a distribution function, and the general answer is in the negative. However, use the uniqueness of the Taylor expansion to show that if a moment generating function $M(s) = Ee^{-sX}$ is convergent for $-\gamma < s < \gamma$ then the distribution of X is uniquely determined by its moments.

20. Write out a proof of the converse part of Theorem 2.3.4.

2.4 TRANSFORMATIONS

It was mentioned in § 1.1.4 that certain functions of random variables are also random variables. If X is a random variable on $(\Omega, \mathscr{F}, P)$ then $Y = g(X)$ is also a random variable provided

$$\{\omega \mid g(X(\omega)) \leqslant y\} \in \mathscr{F}$$

for every real y. This condition on g is not very helpful in practice, and here we are going to restrict attention to those functions that are piecewise 1–1 and differentiable. For example, $g(x) = x^2$, $g(x) = \sin x$, and $g(x) = e^{-sx}$.

Suppose firstly that g is 1–1 (i.e. strictly monotonic) and let g^{-1} denote the function inverse to g. A simple example is

$$y = g(x) = e^x, \qquad -\infty < x < \infty,$$

with inverse

$$x = g^{-1}(y) = \log y, \qquad 0 < y.$$

The monotonicity of g implies that the inequality $a < Y \leqslant b$ is equivalent to $g^{-1}(a) < X \leqslant g^{-1}(b)$ if g is increasing, and to $g^{-1}(b) < X \leqslant g^{-1}(a)$ if g is decreasing. Hence

$$\begin{aligned}\Pr(a < Y \leqslant b) \\ &= \Pr(\min\{g^{-1}(a), g^{-1}(b)\} < X \leqslant \max\{g^{-1}(a), g^{-1}(b)\}).\end{aligned}$$

Introducing the notation

$$F_Y(y) = \Pr(Y \leqslant y), \qquad F_X(x) = \Pr(X \leqslant x),$$

for the respective distribution functions, we write this identity as

$$\begin{aligned}F_Y(b) - F_Y(a) &= F_X(\max\{g^{-1}(a), g^{-1}(b)\}) \\ &\quad - F_X(\min\{g^{-1}(a), g^{-1}(b)\}). \qquad (2.4.1)\end{aligned}$$

Equation (2.4.1) is the basic result concerning 1–1 transformations of random variables.

In the particular case when Y is discrete, say distributed over $y_0, y_1, \ldots,$ the probability distribution $\{\Pr(Y = y_j)\}$ is obtained directly from (2.4.1) on recognising

$$\Pr(Y = y_j) = F_Y(y_j) - F_Y(y_{j-1}), \qquad (2.4.2)$$

and taking $b = y_j$, $a = y_{j-1}$.

In the continuous case we will assume that in addition to monotonicity g is also differentiable. Then the derivatives of both g and g^{-1} do not vanish or change sign. Writing the density functions of X and Y respectively as

$$f_Y(y) = \frac{d}{dy} F_Y(y), \qquad f_X(x) = \frac{d}{dx} F_X(x),$$

formula (2.4.1) becomes

$$\int_a^b f_Y(y)\, dy = \int_{x_0}^{x_1} f_X(x)\, dx,$$

where

$$x_0 = \min\{g^{-1}(a), g^{-1}(b)\}, \qquad x_1 = \max\{g^{-1}(a), g^{-1}(b)\}.$$

We simplify the integral on the right-hand side by the change of variable

$$x = g^{-1}(y).$$

Then

$$dx = \left[\frac{d}{dy} g^{-1}(y)\right] dy$$

and

$$\int_a^b f_Y(y)\, dy = \int_{y_0}^{y_1} f_X(g^{-1}(y)) \frac{d}{dy} g^{-1}(y)\, dy$$

with

$$y_0 = g(\min\{g^{-1}(a), g^{-1}(b)\}), \qquad y_1 = g(\max\{g^{-1}(a), g^{-1}(b)\}).$$

To simplify further suppose firstly that g is increasing. Then g^{-1} is also increasing and

$$\frac{d}{dy}(g^{-1}(y)) > 0, \quad y_0 = g(g^{-1}(a)) = a, \quad y_1 = g(g^{-1}(b)) = b.$$

Hence for increasing g

$$\int_a^b f_Y(y)\, dy = \int_a^b f_X(g^{-1}(y)) \frac{d}{dy} g^{-1}(y)\, dy.$$

On the other hand suppose g is decreasing. Then g^{-1} is also decreasing;

$$\frac{d}{dy}(g^{-1}(y)) < 0, \quad y_0 = g(g^{-1}(b)) = b, \quad y_1 = g(g^{-1}(a)) = a,$$

and

$$\int_a^b f_Y(y)\, dy = -\int_b^a f_X(g^{-1}(y)) \left|\frac{d}{dy} g^{-1}(y)\right| dy$$

$$= \int_a^b f_X(g^{-1}(y)) \left|\frac{d}{dy} g^{-1}(y)\right| dy.$$

Combining the two expressions yields

$$\int_a^b f_Y(y)\, dy = \int_a^b f_X(g^{-1}(y)) \left|\frac{d}{dy} g^{-1}(y)\right| dy. \qquad (2.4.3)$$

Now (2.4.3) holds for all a,b such that $a < b$, and furthermore the integrands on both sides are non-negative. Therefore we have the important expression for the density function of Y:

$$f_Y(y) = f_X(g^{-1}(y)) \left|\frac{d}{dy} g^{-1}(y)\right| \qquad (2.4.4)$$

It is sometimes convenient to write (2.4.4) in a slightly different way. Recall that the derivative dy/dx of $y = g(x)$ is the limit as $\delta x \to 0$ of

$$\frac{\delta y}{\delta x} = \frac{g(x+\delta x)-g(x)}{\delta x},$$

and that $d(g^{-1}(y))/dy$ is the limit as $\delta y \to 0$ of

$$\frac{\delta x}{\delta y} = \frac{g^{-1}(y+\delta y)-g^{-1}(y)}{\delta y}.$$

But

$$\begin{aligned} 1 &= \frac{\delta x}{\delta y}\cdot\frac{\delta y}{\delta x} \\ &= \left[\frac{g^{-1}(y+\delta y)-g^{-1}(y)}{\delta y}\right]\left[\frac{g(x+\delta x)-g(x)}{\delta x}\right] \\ &\to \left[\frac{d}{dy}g^{-1}(y)\right]\left[\frac{d}{dx}g(x)\right] \qquad \text{as } \delta x, \delta y \to 0. \end{aligned}$$

Then

$$\frac{d}{dy}g^{-1}(y) = \{g'(g^{-1}(y))\}^{-1},$$

where

$$g'(g^{-1}(y)) = \frac{d}{dx}g(x)\bigg|_{x=g^{-1}(y)}$$

and (2.4.4) can be rewritten as

$$f_Y(y) = \frac{f_X(g^{-1}(y))}{|g'(g^{-1}(y))|}. \tag{2.4.5}$$

In short:

Theorem 2.4.1 If g is 1–1 and differentiable, the density function of the random variable $Y = g(X)$ is given in terms of the density function of X by (2.4.4) or (2.4.5). If X is discrete then the distribution of Y is given by (2.4.1), (2.4.2) provided only that g is 1–1. □

Example 1 $f_X(x) = \lambda e^{-\lambda x}, \quad 0 < x,$
$\qquad\qquad = 0, \quad 0 \geqslant x.$

Let $Y = g(X) = \sqrt{X}$, $X > 0$, where it is understood that we are taking the positive square root. Then

$$x = g^{-1} = y^2, \qquad \frac{d}{dy}g^{-1}(y) = 2y$$

and

$$f_Y(y) = 2\lambda y e^{-\lambda y^2}, \qquad 0 < y. \qquad \square$$

Example 2 $Y = e^{-sX}$ with s fixed and X an arbitrary continuous random variable. Then

$$x = g^{-1}(y) = -\frac{1}{s}\log y \text{ and } \frac{d}{dy}g^{-1}(y) = -\frac{1}{sy}.$$

The density function of Y (which contains s as a parameter) is

$$f_Y(y) = f_X\left(-\frac{1}{s}\log y\right)\frac{1}{|s|y}, \qquad 0 < y.$$

Note that, as a function of s, the expectation of Y is the moment generating function of X. To verify this:

$$\begin{aligned} EY &= \int_0^\infty y\, f(y)\, dy \\ &= \frac{1}{|s|}\int_0^\infty f_X(-\frac{1}{s}\log y)\, dy \\ &= \int_{-\infty}^\infty e^{-sx} f_X(x)\, dx \end{aligned}$$

on substituting $x = -(\log y)/s$ $\square$

Exercise Work through Example 2 taking

$$f_X(x) = \frac{1}{(2\pi)^{\frac{1}{2}}} e^{-\frac{1}{2}x^2}, \qquad -\infty < x < \infty. \qquad \square$$

The latter half of Example 2 will be recognised as a special case of formula (1.5.1) in which we defined the expectation of a function $g(X)$ of the random variable X as

$$Eg(X) = \int_{-\infty}^\infty g(x) f_X(x)\, dx. \tag{2.4.6}$$

H

The tools are now at hand to give a proof of this expression for those functions g satisfying the conditions of Theorem 2.4.1.

Theorem 2.4.2 If g is 1–1 and differentiable, the expectation $Eg(X)$ is given by (2.4.6).

Proof. Put $Y = g(X)$. By Theorem 2.4.1

$$\begin{aligned} Eg(X) &= \int_{-\infty}^{\infty} y\, f_Y(y)\, dy \\ &= \int_{-\infty}^{\infty} y\, f_X(g^{-1}(y)) \left| \frac{d}{dy} g^{-1}(y) \right| dy \\ &= \int_{-\infty}^{\infty} g(x) f_X(x)\, dx. \end{aligned}$$

The last equality follows from the change of variable $x = g^{-1}(y)$, implying $dx = d(g^{-1}(y))/dy.dy$. □

The class of differentiable functions that are 1–1 excludes too many transformations to be really useful. However we may widen the scope of our results if we agree to consider functions that are only piecewise 1–1 and differentiable. For example, $y = \sin x$, $-\infty < x < \infty$, is not 1–1. However this function is piecewise 1–1 since in each of the intervals $(\frac{1}{2}n\pi, \frac{1}{2}(n+2)\pi)$, $n = \pm 1, \pm 2, \ldots$, there is precisely one solution to the equation $y = \sin x$.

Suppose then that the real line can be partitioned

$$-\infty = a_{-\infty} < \ldots < a_{-1} < a_0 < a_1 < a_2 < \ldots < a_{\infty} = \infty$$

in such a way that $g(x)$ is 1–1 in each of the intervals (a_n, a_{n+1}), $n = 0, \pm 1, \pm 2, \ldots$. The preceding theory is applicable to g(x) written as the sum $\sum_{n=-\infty}^{\infty} g_n(x) = g(x)$ of 1–1 functions

$$\begin{aligned} g_n(x) &= g(x), \quad a_n < x < a_{n+1}, \\ &= 0, \quad \text{otherwise,} \end{aligned}$$

provided we first consider each $g_n(x)$ separately and then sum over n. Writing

$$y = g_n(x), \quad g_n^{-1}(y) = x, \qquad a_n < x < a_{n+1},$$

the contribution to $f_Y(y)$ from x in the interval (a_n, a_{n+1}) is from (2.4.4),

$$f_X(g_n^{-1}(y))\left|\frac{d}{dy}g_n^{-1}(y)\right|.$$

Summing over n yields

$$\begin{aligned} f_Y(y) &= \sum_{n=-\infty}^{\infty} f_X(g_n^{-1}(y))\left|\frac{d}{dy}g_n^{-1}(y)\right| \\ &= \sum_{n=-\infty}^{\infty} \frac{f_X(g_n^{-1}(y))}{|g_n'(g_n^{-1}(y))|}. \end{aligned} \tag{2.4.7}$$

The argument for Theorem 2.4.2 can be extended in the same way and we have

Theorem 2.4.3 If $Y = g(X)$ with g piecewise 1–1 and differentiable then
(a) the density function of Y is given by (2.4.7),
(b) (2.4.6) holds. □

Example 3 $f_X(x) = (2\pi)^{-\frac{1}{2}} e^{-\frac{1}{2}x^2}$, $-\infty < x < \infty$, and

$$Y = g(X) = X^2.$$

The function g is 1–1 separately over $(-\infty, 0)$ and $(0, \infty)$. In the notation used above $g_{-1}^{-1}(y) = -\sqrt{y}$, $g_1^{-1}(y) = +\sqrt{y}$, and

$$\begin{aligned} f_Y(y) &= \frac{f_X(-\sqrt{y})}{2|-\sqrt{y}|} + \frac{f_X(\sqrt{y})}{2|\sqrt{y}|} \\ &= \frac{1}{\sqrt{2\pi}} e^{-\frac{1}{2}y} y^{-\frac{1}{2}} \\ &= \frac{y^{\frac{1}{2}-1} e^{-\frac{1}{2}y}}{\Gamma(\frac{1}{2})2^{\frac{1}{2}}}, \qquad 0 < y. \end{aligned} \tag{2.4.8}$$

The density function (2.4.8) is known either as a *gamma density* with parameter $\frac{1}{2}$ or as a *chi-square* (χ^2) *density* with one degree of freedom (see § 2.5). In the usual terminology we have shown that the square of a standard normal variable is distributed as $\chi^2_{(1)}$, a result that is used frequently in statistical inference. That $\Gamma(\frac{1}{2}) = \sqrt{\pi}$ is also proved in the next section (see 2.5.4). □

Example 4 $Y = \cos X$, with X a continuous random variable on $(-\infty, \infty)$. Then

$$y = g(x) = \cos x, \quad x = g^{-1}(y) = \text{arc}\cos y, \quad \frac{d}{dx}g(x) = -\sin x$$

and using (2.4.7)

$$g'(g^{-1}(y) = -\sin(\text{arc}\cos y) = -(1-y^2)^{\frac{1}{2}}$$

$$f_Y(y) = \frac{1}{(1-y^2)^{\frac{1}{2}}} \sum_{n=-\infty}^{\infty} f_X(g_n^{-1}(y)), \qquad |y| \leqslant 1.$$

In this example there are an infinite number of x satisfying $y = \cos x$ or equivalently an infinite number of y satisfying

$$x = g^{-1}(y) = \text{arc}\cos y.$$

□

Example 5 Let the distribution function of X be $F_X(x) = Pr(X \leqslant x)$ and suppose $F_X(x)$ is strictly increasing. Then if

$$Y = g(X) = F_X(X) \tag{2.4.9}$$

the function g is 1–1, and

$$y = g(x) = F_X(x), \quad x = g^{-1}(y) = F_X^{-1}(y), \quad \frac{dy}{dx} = F'_X(x) = f_X(x).$$

Consequently the density of Y is

$$f_Y(y) = \frac{f_X(F_X^{-1}(y))}{f_X(F_X^{-1}(y))} = 1, \qquad 0 < y < 1.$$

The transformation (2.4.9) yields the uniform density on (0, 1) and is known as the *probability transformation*.

□

Exercises

1. Show that

(a) if $Y = |X|$ then $f_Y(y) = f_X(y) + f_X(-y)$, $0 < y$,
(b) if $Y = aX^2$, $0 < a$, then

$$f_Y(y) = \frac{1}{2(ay)^{\frac{1}{2}}}\left[f_X\left(\sqrt{\frac{y}{a}}\right) + f_X\left(-\sqrt{\frac{y}{a}}\right)\right], \qquad 0 < y,$$

(c) if $Y = \tan X$ then

$$f_Y(y) = (1+y^2)^{-1} \sum_{n=-\infty}^{\infty} f_X(g_n^{-1}(y)), \quad -\infty < y < \infty.$$

Evaluate the right hand members if X is uniformly distributed over $(-\pi, \pi)$.

2. The density function (2.4.8) of a random variable Y distributed as $\chi^2_{(1)}$ was obtained in Example 3 by the transformation $Y = X^2$ with $X \frown N(0,1)$. Show that

$$Ee^{-sY} = (1+2s)^{-\frac{1}{2}},$$

and verify this result by evaluating

$$Ee^{-sX^2} = (2\pi)^{-\frac{1}{2}} \int_{-\infty}^{\infty} e^{-sx^2-\frac{1}{2}x^2}\, dx.$$

In this way the density of functions of a random variable can sometimes be comveniently found by the use of moment generating functions

3. *Random numbers* The probability transformation (Example 5) can be used to artificially generate realisation of a random variable X. Reserving the term 'random number' for a value of y chosen randomly from the interval $(0, 1)$ one has from (2.4.9) the corresponding realisation of X, namely $x = F_X^{-1}(y)$. The random numbers $y_1, y_2, \ldots, y_n$ thus generate the sample of realisations

$$F_X^{-1}(y_j), j = 1, 2, \ldots, n.$$

As a numerical illustration suppose $F_X(x) = 1 - e^{-x}, 0 < x$. Find the sample generated by the random numbers ·48051, ·31695, ·67405, ·87561, and ·13059.

4. Use Example 3 to obtain from tables of both the normal and chi-square distribution the probabilities $\Pr(|X| > 1{\cdot}645)$, $\Pr(|X| > 1{\cdot}96)$, given that $X \frown N(0, 1)$.

5. Follow the reasoning of Example 4 to obtain the general expression for the density of $W = \sin X$.

6. (Continuation.) For fixed s, find the densities of $U_1(s) = \cos sX$ and $V_1(s) = \sin sX$ when X is uniform on $(-\pi, \pi)$. Verify that the expected value of $Z = U_1(s)+iV_1(s)$ is the characteristic function of X.

2.5 SPECIAL FUNCTIONS

This section discusses briefly some special functions that arise frequently in probability theory. It is divided into three sub-sections (i) gamma and beta functions, (ii) binomial coefficients and the hypergeometric function, and (iii) indicator functions. Finally there is a table of the more commonly used generating functions.

(i) Gamma and beta functions

The gamma function $\Gamma(\alpha)$ is defined by the integral

$$\Gamma(\alpha) = \int_0^\infty e^{-x}x^{\alpha-1}\,dx \tag{2.5.1}$$

which is convergent for $\alpha > 0$. For $\alpha > 1$ an integration by parts gives

$$\Gamma(\alpha) = (\alpha-1)\int_0^\infty e^{-x}x^{\alpha-2}\,dx = (\alpha-1)\Gamma(\alpha-1), \tag{2.5.2}$$

and, if α is an integer, repeated application of (2.5.2) shows that

$$\Gamma(\alpha) = (\alpha-1)(\alpha-2)\ldots 2.1.\Gamma(1).$$

But $\Gamma(1) = \int_0^\infty e^{-x}\,dx = 1$. Hence for any positive integer n

$$\Gamma(n) = (n-1)!. \tag{2.5.3}$$

The gamma function (2.5.1) may therefore be correctly described as a generalisation of the factorial function. Note in particular that

$$0! = \Gamma(1) = 1.$$

In applications $\Gamma(\alpha)$ commonly appears either with α an integer (in which case (2.5.3) applies) or with $\alpha = \frac{1}{2}n$ for integer n. In the latter case using (2.5.2) again,

$$\Gamma(\tfrac{1}{2}n) = (\tfrac{1}{2}n-1)(\tfrac{1}{2}n-2)\ldots\tfrac{1}{2}.\Gamma(\tfrac{1}{2}).$$

But from the definition (2.5.1)

$$\{\Gamma(\tfrac{1}{2})\}^2 = \int_0^\infty \int_0^\infty (xy)^{-\frac{1}{2}} e^{-(x+y)}\, dx dy$$

$$= 4\int_0^\infty \int_0^\infty e^{-(u^2+v^2)}\, du dv \qquad \text{(on putting } x = u^2, y = v^2\text{)}.$$

Next, putting $u = r\sin\theta$, $v = r\cos\theta$ to find $dudv = r\,drd\theta$, we get

$$\{\Gamma(\tfrac{1}{2})\}^2 = 4\int_0^\infty \int_0^{\frac{1}{2}\pi} re^{-r^2}\, drd\theta = 4\int_0^{\frac{1}{2}\pi} \tfrac{1}{2}\, d\theta = \pi.$$

Hence*

$$\Gamma(\tfrac{1}{2}) = \sqrt{\pi}. \tag{2.5.4}$$

The $\chi^2_{(1)}$ density function obtained as the density of the square of a standard normal variable in Example 3 of the preceding section is a special case of the family of *gamma probability density functions.* From (2.5.1)

$$f(x) = \frac{e^{-x}x^{\alpha-1}}{\Gamma(\alpha)}, \qquad 0 < x, \tag{2.5.5}$$

is a density function and is called a *gamma density with parameter* α. Random variables with this density are said to be distributed according to the gamma law with parameter α. The special case when $\alpha = \frac{1}{2}n$, n an integer, is the *chi-square density* with n degrees of freedom (abbreviated $\chi^2_{(n)}$) and it is customary to write

$$f(x) = \frac{e^{-\frac{1}{2}x}x^{\frac{1}{2}n-1}}{\Gamma(\frac{1}{2}n)2^{\frac{1}{2}n}}, \qquad 0 < x. \tag{2.5.6}$$

The beta function $B(\alpha, \beta)$ is defined by

$$B(\alpha, \beta) = \int_0^1 x^{\alpha-1}(1-x)^{\beta-1}\, dx \tag{2.5.7}$$

for $\alpha, \beta > 0$. An alternative representation is

$$B(\alpha, \beta) = \frac{\Gamma(\alpha)\Gamma(\beta)}{\Gamma(\alpha+\beta)} \tag{2.5.8}$$

* Precisely the same argument shows that $(2\pi)^{-\frac{1}{2}} \int_{-\infty}^{\infty} e^{-\frac{1}{2}x^2}\, dx = 1$. We find $\int_0^\infty e^{-\frac{1}{2}x^2}\, dx = (\frac{1}{2}\pi)^{\frac{1}{2}}$ and the result follows by symmetry (See Exercise 4).

obtained in the following way:

$$\Gamma(\alpha)\Gamma(\beta) = \int_0^\infty \left[\int_0^\infty e^{-(x+y)}x^{\alpha-1}y^{\beta-1}\,dx\right]dy.$$

Keeping y fixed, put $u = x/(x+y)$ to give $x = uy/(1-u)$ and

$$dx = \frac{ydu}{(1-u)} + \frac{uydu}{(1-u)^2} = \frac{ydu}{(1-u)^2}.$$

Then

$$\Gamma(\alpha)\Gamma(\beta) = \int_0^\infty \left[\int_0^1 \left\{exp\left(\frac{-y}{1-u}\right)\right\}\left(\frac{u}{1-u}\right)^{\alpha-1} y^{\alpha+\beta-1}\,du\right]dy$$

$$= \int_0^1 \left[\int_0^\infty \left\{exp\left(\frac{-y}{1-u}\right)\right\} y^{\alpha+\beta-1}\,dy\right]\left[\frac{u}{1-u}\right]^{\alpha-1} du$$

$$= \int_0^1 \left[\int_0^\infty e^{-v}v^{\alpha+\beta-1}\,dv\right]u^{\alpha-1}(1-u)^{\beta-1}\,du$$

(putting $y = (1-u)v$)

$$= \Gamma(\alpha+\beta)\int_0^1 u^{\alpha-1}(1-u)^{\beta-1}\,du.$$

Hence (2.5.8).

The *beta distribution* is defined by the density function

$$f(x) = \frac{1}{B(\alpha, \beta)}x^{\alpha-1}(1-x)^{\beta-1}, \qquad 0 < x < 1. \tag{2.5.9}$$

The change of variable $y = \beta x/\alpha(1-x)$, with

$$x = \frac{\alpha y}{\beta+\alpha y}, \qquad dx = \frac{\alpha\beta}{(\beta+\alpha y)^2}dy,$$

gives an alternative form for the density;

$$f(y) = \frac{1}{B(\alpha,\beta)}\left[\frac{a}{\beta}\right]^{\alpha}\frac{y^{\alpha-1}}{(1+\alpha y/\beta)^{\alpha+\beta}}, \qquad 0 < y < \infty. \tag{2.5.10}$$

Special cases are:

(a) *Uniform or rectangular distribution* on (0, 1). Take $\alpha = \beta = 1$ in (2.5.9) to find

$$f(x) = 1, \qquad 0 < x < 1. \tag{2.5.11}$$

(b) *Arcsine distribution.* If $\alpha = \beta = \frac{1}{2}$ in (2.5.9)

$$f(x) = \frac{1}{\pi(x(1-x))^{\frac{1}{2}}}, \qquad 0 < x < 1. \tag{2.5.12}$$

The name arises from the fact that the distribution function is

$$F(x) = \frac{2}{\pi} \arcsin \sqrt{x}, \qquad 0 < x < 1.$$

(c) The F distribution with (m, n) degrees of freedom (abbreviated $F_{(m,n)}$) has density (2.5.10) with $\alpha = \frac{1}{2}m$, $\beta = \frac{1}{2}n$. It is discussed further in § 4.4.

(ii) Binomial coefficients and the hypergeometric function

The binomial coefficient $\binom{N}{n}$ appeared in § 1.3 as the number of ways of selecting n objects from N when sampling without replacement. The definition of $\binom{N}{n}$ for both N, n positive integers was

$$\binom{N}{n} = \frac{(N)_n}{n!} = \frac{N(N-1)\ldots(N-n+1)}{n!}$$

and we agree now to extend this definition to include the case when N is any real number. Thus for all real ν and positive integer n

$$\binom{\nu}{n} = \frac{(\nu)_n}{n!}, \qquad n = 1, 2, 3, \ldots,$$

$\binom{\nu}{0} = 1$, and $\binom{\nu}{n}$ is not defined for other values of n. The generating function of the binomial coefficients is

$$\sum_{n=0}^{\infty} (\nu)_n z_n/n! = \sum_{n=0}^{\infty} \binom{\nu}{n} z^n = (1+z)^\nu,$$

being a terminating series if ν is a positive integer.

More generally consider the series

$${}_2F_1(\alpha, \beta; \gamma; z) = \sum_{j=0}^{\infty} \frac{(-\alpha)_j(-\beta)_j(-z)^j}{(-\gamma)_j j!}$$

which is called the *hypergeometric function* with parameters α, β, γ. Writing

$$[\alpha]_j = \alpha(\alpha+1)\dots(\alpha+j-1) = (-1)^j(-\alpha)_j,$$

we obtain the series by which the hypergeometric function is usually defined

$$_2F_1(\alpha, \beta; \gamma; z) = \sum_{j=0}^{\infty} \frac{[\alpha]_j [\beta]_j z^j}{[\gamma]_j j!}. \tag{2.5.13}$$

Hypergeometric functions are of some importance in applied mathematics and are solutions of the differential equation

$$z(1-z)\frac{d^2y}{dz^2} + \{\gamma-(1+\alpha+\beta)z\}\frac{dy}{dz} - \alpha\beta y = 0. \tag{2.5.14}$$

Assume that γ is not zero or a negative integer. Then (2.5.13) can be either a finite or infinite series. It is a finite series if either α or β are negative integers, say $\alpha = -n$, for then

$$[\alpha]_{n+1} = [-n]_{n+1} = (-n)(-n+1)\dots(-1)\,0 = 0$$

in which case $_2F_1(-n, \beta; \gamma; z)$ is a polynomial in z of degree n. A case of special interest is

$$\begin{aligned} _2F_1(-n, \beta; \beta; -z) &= \sum_{j=0}^{n} [-n]_j \frac{(-z)^j}{j!} \\ &= \sum_{j=0}^{n} \binom{n}{j} z^j = (1+z)^n. \end{aligned}$$

It follows that the probability generating function of the binomial distribution can be written as

$$\begin{aligned} \sum_{n=0}^{n} \binom{n}{j} p^j(1-p)^{n-j}z^j &= (1-p+pz)^n \\ &= (1-p)^n \left[1+\frac{pz}{1-p}\right]^n \\ &= (1-p)^n \, {}_2F_1\left(-n, \beta; \beta; \frac{-pz}{1-p}\right). \end{aligned}$$

The probability distribution defined in (1.3.5)

$$h_j = \binom{M}{j}\binom{N-M}{n-j}\bigg/\binom{N}{n}, \qquad j = 0, 1, \dots, \min(n, M).$$

was called the *hypergeometric distribution*. The name arises because h_j can be rewritten in terms of the coefficients of a finite hypergeometric series. After some manipulation we find

$$h_j = \frac{(N-M)_n [-n]_j [-M]_j}{(N)_n [N-M-n+1]_j} \cdot \frac{1}{j!},$$

with probability generating function

$$\sum_{j=0}^{\min(n,M)} z^j h_j = \frac{(N-M)_n}{(N)_n} {}_2F_1(-n, -M; N-M-n+1; z).$$

From Exercise 1.3.7 it is known that $\sum_j h_j = 1$. Moments of the distribution can be obtained from (2.5.14). For example, substituting $z = 1$ in the differential equation with appropriate values for α, β, γ gives

$$EX = \sum_j jh_j = nM/N.$$

Higher moments follow on differentiating (2.5.14).

This procedure can be repeated whenever a probability generating function can be written as a hypergeometric series, as is the case with many commonly occurring discrete distributions. Another example is the geometric distribution for which

$$\sum_{j=0}^{\infty} (1-\rho)\rho^j z^j = \frac{1-\rho}{1-\rho z} = (1-\rho)\, {}_2F_1(1, \beta; \beta; \rho z).$$

We do not make much use of these results. It is interesting to note, however, that a partial unification of the theory for discrete distributions can be achieved through the hypergeometric function (2.5.13). In particular, if the parameters α, β, γ are such that $y = y(z)$ satisfying (2.5.14) is a probability generating function, then successive differentiation of this equation at $z = 1$ yields a recurrence relation for moments.

Finally we recall *Stirling's formula* (Exercise 1.3.14)

$$n! \sim (2\pi)^{\frac{1}{2}} n^{n+\frac{1}{2}} e^{-n}, \qquad n \to \infty, \tag{2.5.15}$$

used to approximate the factorials of large numbers.

(iii) Indicator functions

Suppose $(\Omega, \mathscr{F}, P)$ is a probability space and that A is an event, that is $A \in \mathscr{F}$ is a subset of Ω with probability $P(A)$. The *indicator function* I_A of A is defined to be

$$\begin{aligned} I_A(\omega) = I_A &= 1 \quad \text{if } \omega \in A, \\ &= 0 \quad \text{if } \omega \notin A. \end{aligned} \tag{2.5.16}$$

According to Definitions 1.4.1 and 1.4.3, for each $A \in \mathscr{F}$, I_A is a random variable. Its distribution is

$$\Pr(I_A = 1) = P(A), \qquad \Pr(I_A = 0) = 1 - P(A), \tag{2.5.17}$$

with expectation

$$EI_A = P(A). \tag{2.5.18}$$

Examples of indicator functions have appeared in § 1.4.

The usefulness of indicator functions lies partly in the fact that they translate set operations into arithmetic ones, with intersection corresponding to multiplication and the union of disjoint events to addition. Thus for $A, B \in \mathscr{F}$,

$$I_{AB} = I_A I_B$$

(since $I_{AB} = 1$ if and only if $\omega \in AB$). If A and B are disjoint then (since ω can be a member of at most one of A and B)

$$I_{A \cup B} = I_A + I_B.$$

By induction one concludes that if $A_1, A_2, \ldots, A_k \in \mathscr{F}$ then

$$I_{\bigcap_{j=1}^{k} A_j} = \prod_{j=1}^{k} I_{A_j},$$

and for disjoint sets

$$I_{\bigcup_{j=1}^{k} A_j} = \sum_{j=1}^{k} I_{A_j},$$

Note also that $I_A = 1 - I_{\bar{A}}$. Hence for not necessarily disjoint A, B

$$I_{\bar{A}B} = I_{\bar{A}} I_B = (1 - I_A) I_B$$

and we have the general formula concerning unions (c.f. (1.2.5))

$$\begin{aligned} I_{A \cup B} = I_{A \cup \bar{A}B} &= I_A + I_{\bar{A}B} \\ &= I_A + I_B - I_A I_B. \end{aligned}$$

These formulae are useful in certain combinatorial problems.

The main theoretical use of indicator functions is that they make possible a different development of the notions of random variable and expectation to that put forward in § 1.4 and § 1.5. To outline this approach suppose for a given sample space $(\Omega, \mathscr{F}, P)$ we take (2.5.16)–(2.5.18) as *defining* what is meant by a zero-one random variable (i.e. indicator function) and by what is meant by expectation. Then any linear combination of indicator functions

$$X = \sum_{j=1}^{n} a_j I_{A_j}$$

is called a *simple random variable*, the a_j being (finite) constants. The expectation EX of a simple random variable is *defined* to be

$$EX = \sum_{j=1}^{n} a_j EI_{A_j} = \sum_{j=1}^{n} a_j P(A_j)$$

The notion of a simple random variable can clearly be extended to countable sample spaces with $X = \sum_{j=-\infty}^{\infty} a_j I_{A_j}$, and indeed to the non-denumerable case. However, it is not the intention to pursue this topic but to note only the way in which this use of indicator functions makes more reasonable the apparently arbitrary definitions in § 1.4 of random variables as functions from the sample space into $(-\infty, \infty)$. Indicator functions, taking only the values zero or one, are the simplest such functions, and more complicated random variables can be regarded as their generalisations and extensions.

Exercises

1. Show that the convolution of the gamma density functions $\gamma_\alpha(x)$, $\gamma_\beta(x)$ is $\gamma_{\alpha+\beta}(x)$ by

   (a) using moment generating functions,
   (b) direct integration.

2. If

$$\Gamma_k(x) = \frac{1}{(k-1)!}\int_0^x e^{-y} y^{k-1}\, dy$$

is the gamma–k distribution function with k a positive integer, use repeated integration by parts to show that

$$\Gamma_k(x) = \sum_{j=k}^{\infty} e^{-x}x^j/j!.$$

Hence if X_k is gamma–k and Y_x is Poisson with mean x,

$$\Pr(X_k \leqslant x) = \Pr(Y_x \geqslant k).$$

Numerical values of $\Gamma_k(x)$ can be found from tables of the Poisson distribution and vice-versa.

3. Find the mean and variance of the gamma and beta distributions.

4. Referring to the footnote on page 99, write out in detail the proof that $(2\pi)^{-\frac{1}{2}} \int_{-\infty}^{\infty} e^{-\frac{1}{2}x^2}\, dx = 1$.

5. If $\phi(x) = (2\pi)^{-\frac{1}{2}}\, e^{-\frac{1}{2}x^2}$ is the standard normal density and $\Phi(x) = \int_{-\infty}^{x} \phi(y)\, dy$, show that for $x > 0$

$$\frac{1}{x}\left(1-\frac{1}{x^2}\right)\phi(x) < 1-\Phi(x) < \frac{1}{x}\,\phi(x)$$

[FELLER I, page 175].

6. Verify

(a) $$\binom{n}{j-1}+\binom{n}{j}=\binom{n+1}{j},$$

(b) $$\binom{-n}{j}=\binom{n+j-1}{j}(-1)^j,$$

(c) $$\sum_{j=0}^{n}\binom{a}{j}\binom{b}{n-j}=\binom{a+b}{n}$$

(This is equivalent to proving that the hypergeometric probabilities sum to unity),

GENERATING FUNCTIONS (continued)

Name	$f(x)$	Ee^{-sX}
Uniform	$1, \quad 0 < x < 1$	$\dfrac{1-e^{-s}}{s}, \quad -\infty < s < \infty$
Uniform	$1/2a, \quad -a < x < a$	$\dfrac{1}{2as}(e^{as}-e^{-as}) = \dfrac{\sinh as}{as}, \quad \lvert s\rvert < \infty$
Bessel*	$\alpha e^{-x} I_\alpha(x)/x, \quad 0 < x$	$[1+s-\{(1+s)^2-1\}^{\frac{1}{2}}]^\alpha, \quad 0 < s$
Two-sided exponential	$\frac{1}{2}e^{-\lvert x\rvert}, \quad -\infty < x < \infty$	$1/(1-s^2), \quad \lvert s\rvert < 1$
Triangular	$\left(1-\dfrac{\lvert x\rvert}{a}\right)\dfrac{1}{a}, \quad \lvert x\rvert < a$	$\dfrac{2(\cosh as - 1)}{a^2 s^2}, \quad \lvert s\rvert < \infty$
—	$\left(\dfrac{2}{\pi}\right)^{\frac{1}{2}} e^{-\frac{1}{2}x^2}, \quad 0 < x$	$\dfrac{2e^{\frac{1}{2}s^2}}{(2\pi)^{\frac{1}{2}}} \displaystyle\int_s^\infty e^{-\frac{1}{2}y^2}\,dy, \quad \lvert s\rvert < \infty$
—	$\dfrac{e^{-1/(2x)}}{(2\pi x^3)^{\frac{1}{2}}}, \quad 0 < x$	$e^{-(2s)^{1/2}}, \quad 0 < s.$

* $I_\alpha(x) = (x^\alpha/2^\alpha\Gamma(\alpha+1)) \sum_{j=0}^{\infty} (x^2/4)^j/j![\alpha+1]_j$ defines one variety of Bessel function.

GENERATING FUNCTIONS (*continued*)

Name	$f(x)$	Ee^{isX}
Cauchy	$\dfrac{1}{\pi\{1+(x-a)^2\}}, \quad -\infty < x < \infty$	$e^{ias}\,e^{-\lvert s\rvert}, \quad \lvert s\rvert < \infty$
Triangular	$\left(1-\dfrac{\lvert x\rvert}{a}\right)\dfrac{1}{a}, \quad \lvert x\rvert < a$	$\dfrac{2(1-\cos as)}{a^2 s^2}, \quad \lvert s\rvert < \infty$
—	$\dfrac{1-\cos ax}{\pi a x^2}, \quad -\infty < x < \infty$	$1-\dfrac{\lvert s\rvert}{a}, \quad \lvert s\rvert \leqslant a$ $0 \quad , \quad \lvert s\rvert > a$

Name	Joint density	$E\exp\left\{\prod_{j=1}^{k} s_j X_j\right\}$
Standard bivariate normal	$\dfrac{\exp[-\frac{1}{2}(x_1^2 - 2\rho x_1 x_2 + x_2^2)/(1-\rho^2)]}{2\pi(1-\rho^2)^{1/2}}, \quad -\infty < x_1, x_2 < \infty$	$\exp[\frac{1}{2}(s_1^2 + 2\rho s_1 s_2 + s_2^2)], \quad -\infty < s_1,$
Multinomial	$\dfrac{n!\prod_{j=1}^{k+1} p_j^{n_j}}{\prod_{j=1}^{k+1} n_j!}, \quad \sum_{j=1}^{k+1} n_j = n, \quad \sum_{j=1}^{k+1} p_j = 1$	$\left\{1-\sum_{j=1}^{k}(1-e^{-s_j})\,p_j\right\}^n, \quad -\infty < s_1, \ldots, s_k < \infty$

Characteristic functions are obtained from moment generating functions on replacing $-s$ by is.

Chapter 3
SEVERAL RANDOM VARIABLES

The basic concepts of probability theory and some ancillary mathematical techniques have been introduced in the two preceding chapters. We proceed to discuss the joint behaviour of several random variables. That is, we have the functions

$$X_1(\omega), X_2(\omega), \ldots, X_n(\omega)$$

from a sample space $\Omega = \{\omega\}$ into the real line and in § 3.1 we extend the notion of distribution function and associated quantities to this more general case.

The important sections of the chapter are the first three. § 3.4 and § 3.5 are in the nature of a digression, since whilst they treat material important in developments of the subject, they are not necessary for the further basic theory in Chapters 4 and 5. On the other hand the notions of conditional probability and independence (§§ 3.2, 3.3) are of crucial importance and the two subsequent chapters deal largely with properties of independent random variables.

3.1 JOINT DISTRIBUTIONS

Let X and Y be random variables defined on the probability space $(\Omega, \mathscr{F}, P)$. That is, $X = X(\omega)$ and $Y = Y(\omega)$ are two functions from Ω into $(-\infty, \infty)$. Our present concern is to extend the probabilistic statements about a single random variable discussed in Chapter 1 to similar statements about two (or more) random variables considered simultaneously. Recall from § 1.4 that the probability that X realises a value in $(x_0, x]$ is

$$\Pr(x_0 < X \leqslant x) = P(\{\omega \mid x_0 < X \leqslant x\}).$$

Considering X and Y jointly the probability that simultaneously one has $x_0 < X \leqslant x$ and $y_0 < Y \leqslant y$ is, by extension,

$$\begin{aligned}\Pr(x_0 < X \leqslant x, y_0 < Y \leqslant y) \\ &= P(\{\omega \mid x_0 < X \leqslant x, y_0 < Y \leqslant y\}) \\ &= P(\{\omega \mid x_0 < X \leqslant x\} \cap \{\omega \mid y_0 < Y \leqslant y\}). \qquad (3.1.1)\end{aligned}$$

In words, the probability of the joint realisation of the two inequalities is the P value of those sample points for which both inequalities are satisfied, i.e. the P value assigned to the intersection $\{\omega \mid x_0 < X \leqslant x\} \cap \{\omega \mid y_0 < Y \leqslant y\}$.

If Ω is discrete, say the integers, the substitution $x_0 = x-1$, $y_0 = y-1$, in (3.1.1) gives

$$p_{x,y} = \Pr(X = x, Y = y) = P(\{\omega \mid X = x, Y = y\}). \qquad (3.1.2)$$

Considered as a function of the two discrete variables x and y, $\{p_{x,y}\}$ defined by (3.1.2) is called the *joint probability distribution* of the (discrete) random variables X and Y.

Example 1 Two dice are rolled, $\Omega = \{(a, b) \mid a, b = 1, 2, \ldots, 6\}$. Let X denote the sum of the two faces and Y the absolute value of their difference. X is distributed over the integers $2, 3, \ldots, 12$, and Y over $0, 1, \ldots, 5$. Then, for example,

$$\begin{aligned}\Pr(X = 4, Y = 2) &= P(\{\omega \mid X = 4, Y = 2\}) \\ &= P(\{(a, b) \mid a+b = 4, |a-b| = 2\}) \\ &= P(\{(3, 1)\} \cup \{(1, 3)\}) \\ &= P(\{(3, 1)\}) + P(1, 3)\}),\end{aligned}$$

which is 1/18 if the dice are fair. □

Whether or not Ω is discrete the choice $x_0 = y_0 = -\infty$ in (3.1.1) gives for $-\infty < x, y < \infty$,

$$F(x, y) = \Pr(X \leqslant x, Y \leqslant y) = P(\{\omega \mid X \leqslant x, Y \leqslant y\}), \qquad (3.1.3)$$

called the *joint probability distribution function of X and Y*. $F(x, y)$ may be a step function (if X and Y are discrete), continuous in both variables (if X and Y are continuous), or a mixture of the two.

In any case $F(x, y)$ is non-decreasing in both x and y. To see this suppose $x_1 \geqslant x, y_1 \geqslant y$. Then

$$\{\omega \mid X \leqslant x, Y \leqslant y\} \subseteq \begin{Bmatrix} \{\omega \mid X \leqslant x_1, Y \leqslant y\} \\ \{\omega \mid X \leqslant x, \ Y \leqslant y_1\} \end{Bmatrix}$$
$$\subseteq \{\omega \mid X \leqslant x_1, Y \leqslant y_1\},$$

and since $P(A) \leqslant P(B)$ if $A \subseteq B$ (equation 1.2.4),

$$F(x, y) \leqslant \begin{Bmatrix} F(x_1, y) \\ F(x, y_1) \end{Bmatrix} \leqslant F(x_1, y_1).$$

As in § 1.4 zero probability is assigned to the event

$$\lim_{x, y \to -\infty} \{\omega \mid X \leqslant x, Y \leqslant y\}.$$

Hence $\lim\limits_{x, y \to -\infty} F(x, y) = 0$, and $F(x, y)$ is non-decreasing from zero as x, y vary from $-\infty$ to ∞.

It turns out that $\lim\limits_{x, y \to \infty} F(x, y) = 1$ but before reaching this point it is convenient to introduce another term. Letting $y \to \infty$ in (3.1.3) we have* by the theorem of the Appendix to Chapter 1

$$F_X(x) = \lim_{y \to \infty} F(x, y) = P(\{\omega \mid X \leqslant x, Y < \infty\})$$
$$= P(\{\omega \mid X \leqslant x\}), \qquad (3.1.4)$$

which is the distribution function of X. In the context of joint distributions $F_X(x)$ is called the *marginal distribution function of X*. Similarly

$$F_Y(y) = \lim_{x \to \infty} F(x, y) = P(\{\omega \mid Y \leqslant y\})$$

is the marginal distribution function of Y. Since $F_X(x)$ is a distribution function increasing to unity as $x \to \infty$ it follows that

$$\lim_{x, y \to \infty} F(x, y) = 1$$

* The argument is essentially the same as on page 10. Thus

$$\lim_{y \to \infty} P(\{\omega \mid X \leqslant x, Y \leqslant y\}) = P(\lim_{y \to \infty} \{\omega \mid X \leqslant x\} \cap \{\omega \mid Y \leqslant y\})$$
$$= P(\{\omega \mid X \leqslant x\} \cap \Omega) = P(\{\omega \mid X \leqslant x\}).$$

If X and Y are continuous and $F(x, y)$ is differentiable in both variables, the derivative

$$\frac{\partial^2}{\partial x \partial y} F(x, y) = f(x, y) \tag{3.1.5}$$

is called the *joint probability density function of X and Y*. As in the univariate case it is interpreted by

$$\Pr(x < X \leqslant x+dx, y < Y \leqslant y+dy) = f(x, y)\, dxdy + o(dxdy),$$

a result that can be verified from the bivariate Taylor expansion of $F(x, y)$. The *marginal density function of X is*

$$\begin{aligned} f_X(x) &= \int_{-\infty}^{\infty} f(x, y)\, dy = \frac{d}{dx} F(x, \infty) \\ &= \frac{d}{dx} F_X(x). \end{aligned} \tag{3.1.6}$$

The marginal density of Y is obtained in the same way.

Having defined the joint distribution of X and Y it is natural to introduce joint generating functions. In the discrete non-negative case

$$E z_1^X z_2^Y = \sum_x \sum_y z_1^x z_2^y \Pr(X = x, Y = y)$$

is the *joint probability generating function*, convergent for at least $|z_1| \leqslant 1, |z_2| \leqslant 1$, and in the continuous case

$$E e^{-(s_1 X + s_2 Y)} = \int_{-\infty}^{\infty} \int_{-\infty}^{\infty} e^{-(s_1 x + s_2 y)} f(x, y)\, dxdy$$

is the *joint moment generating function* (provided the integral on the right is convergent). More generally $E e^{i(s_1 X + s_2 Y)}$, s_1, s_2 real, is the *joint characteristic function of X and Y*.

In addition to the moments EX^k, EY^j, computed from the respective marginal distributions, one also has *product* or *joint moments*, namely $E(X^k Y^j)$. In the continuous case

$$E(X^k Y^j) = \int_{-\infty}^{\infty} \int_{-\infty}^{\infty} x^k y^j f(x, y)\, dxdy \tag{3.1.7}$$

if the integral on the right converges absolutely. An analogous definition holds for discrete variables. Of particular interest is the first joint moment

$$E(XY) = \int\int xy\, f(x, y)\, dxdy.$$

It is apparent that in general $E(XY) \neq (EX)(EY)$ and the difference of these two quantities is called the *covariance of X and Y*, written

$$\mathrm{Cov}(X, Y) = E(XY) - (EX)(EY).$$

The covariance of X and Y is usually defined by

$$\mathrm{Cov}(X, Y) = E((X - EX)(Y - EY)). \tag{3.1.8}$$

However, expansion of the product on the right hand side recovers the first version. The name covariance arises from the fact that (3.1.8) is an extension of the notion of variance, since writing $Y = X$ in (3.1.7) yields

$$\mathrm{Var}\, X = \mathrm{Cov}(X, X) = E(X - EX)^2.$$

The quantity

$$\mathrm{Corr}(X, Y) = \frac{\mathrm{Cov}(X, Y)}{((\mathrm{Var}\, X)(\mathrm{Var}\, Y))^{\frac{1}{2}}} \tag{3.1.9}$$

is called the *correlation coefficient of X and Y*.

A common notation for the correlation coefficient is

$$\rho_{X,Y} = \mathrm{Corr}(X, Y).$$

In Exercise 6 you will be asked to show that $-1 \leqslant \rho_{X,Y} \leqslant 1$ for any pair of random variables. Note that $\rho_{X,Y} = 0$ if and only if the covariance vanishes i.e. $E(XY) = (EX)(EY)$. Random variables with zero correlation coefficient are said to be *orthogonal.*

Example 2 The following density defines the *standard bivariate normal distribution*:

$$f(x, y) = \frac{1}{2\pi(1-\rho^2)^{\frac{1}{2}}} \exp\left[-\frac{(x^2 - 2\rho xy + y^2)}{2(1-\rho^2)}\right],$$

$$|\rho| < 1, \; -\infty < x, y < \infty$$

We first find $f_X(x)$, the marginal density of X.

$$f_X(x) = \frac{1}{2\pi(1-\rho^2)^{\frac{1}{2}}}\int_{-\infty}^{\infty} \exp\left[-\frac{(x^2-2\rho xy+y^2)}{2(1-\rho^2)}\right] dy$$

$$= \frac{\exp(-x^2/2(1-\rho^2))}{2\pi(1-\rho^2)^{\frac{1}{2}}}\int_{-\infty}^{\infty} \exp\left[-\frac{((y-\rho x)^2-\rho^2x^2)}{2(1-\rho^2)}\right] dy$$

$$= \frac{\exp(-(x^2-\rho^2x^2)/2(1-\rho^2))}{(2\pi)^{\frac{1}{2}}(1-\rho^2)^{\frac{1}{2}}}\int_{-\infty}^{\infty} \exp\left[-\frac{(y-\rho x)^2}{2(1-\rho^2)}\right] dy$$

$$= \frac{\exp(-\frac{1}{2}x^2)}{(2\pi)^{\frac{1}{2}}}, \qquad -\infty < x < \infty.$$

Thus $X \frown N(0, 1)$. Repeating the argument (or by symmetry) we find also that $Y \frown N(0, 1)$. Next consider

$$E(XY) = \frac{1}{2\pi(1-\rho^2)^{\frac{1}{2}}}\int_{-\infty}^{\infty}\int_{-\infty}^{\infty} xy \exp\left[-\frac{(x^2-2\rho xy+y^2)}{2(1-\rho^2)}\right] dxdy,$$

which after completing the square in the exponent and integrating successively reduces to

$$E(XY) = \rho.$$

Since X and Y both have zero means and unit variance, it follows from (3.1.9) that the correlation between X and Y is

$$\text{Corr}(X, Y) = \rho.$$

Finally, as an exercise, you are asked to show that the joint moment generating function is

$$Ee^{-(s_1X+s_2Y)} = e^{\frac{1}{2}(s_1^2+2\rho s_1s_2+s_2^2)},$$

and to verify that

$$\left.\frac{\partial^2 Ee^{-(s_1X+s_2Y)}}{\partial s_1 \partial s_2}\right|_{\substack{s_1=0\\ s_2=0}} = E(XY) = \rho. \qquad \square$$

Clearly the preceding remarks can be extended to define the

joint distribution of the random variables $X_1, X_2, \ldots, X_n$ on $(\Omega, \mathcal{F}, P)$ by the distribution function

$$F(x_1, x_2, \ldots, x_n) = \Pr(X_1 \leqslant x_1, X_2 \leqslant x_2, \ldots, X_n \leqslant x_n)$$
$$= P(\{\omega \mid X_1 \leqslant x_1, X_2 \leqslant x_2, \ldots, X_n \leqslant x_n\}),$$
$$-\infty < x_1, x_2, \ldots, x_n < \infty. \qquad (3.1.10)$$

Note that for a function of $x_1, x_2, \ldots, x_n$ to be a distribution function it must be non-decreasing from 0 to 1 as each of the n variables increase from $-\infty$ to ∞. Furthermore for each j, $1 \leqslant j \leqslant n$,

$$\lim_{x_j \to -\infty} F(x_1, x_2, \ldots, x_n) = 0,$$

and

$$\lim_{x_j \to \infty} F(x_1, x_2, \ldots, x_n)$$
$$= F(x_1 \ldots, x_{j-1}, \infty, x_{j+1}, \ldots, x_n)$$
$$= F_{X_1, \ldots X_{j-1}, X_{j+1}, \ldots, X_n}(x_1, \ldots, x_{j-1}, x_{j+1}, \ldots, x_n)$$

must also be a distribution function, called the *marginal distribution function of* $X_1, \ldots, X_{j-1}, X_{j+1}, \ldots, X_n$. Repeating this procedure one finally retrieves the marginal distribution function of a single random variable, say X_k, as

$$F_{X_k}(\infty, \ldots, \infty, x_k, \infty, \ldots, \infty) = P(\{\omega \mid X_k \leqslant x_k\}).$$

At this stage it is convenient to introduce a specialisation of the sample space $\Omega = \{\omega\}$. Considering only the two random variables X_1 and X_2 for the moment suppose that X_1 is defined as a function on $\Omega_1 = \{w_1\}$ and X_2 on $\Omega_2 = \{w_2\}$. Simultaneous discussion of X_1 and X_2 requires the introduction of the product sample space

$$\Omega = \Omega_1 \times \Omega_2 = \{\omega \mid \omega = (w_1, w_2), w_1 \in \Omega_1, w_2 \in \Omega_2\}.$$

The elements of Ω are two-dimensional vectors, the first coordinate taking values in the domain of X_1 and the second in the domain of X_2. On this sample space we have the *vector* random variable (X_1, X_2) which is a function from Ω into the plane $(-\infty, \infty) \times (-\infty, \infty)$.

Example 3 A die and a coin are tossed simultaneously. Let X_1 be the random variable denoting the face shown by the die and let X_2 correspond to that shown by the coin. Then $\Omega_1 = \{1, 2, 3, 4, 5, 6\}$, $\Omega_2 = \{H, T\}$, and the product space on which (X_1, X_2) is defined is

$$\Omega = \{(w_1, w_2) \mid w_1 = 1, 2, \ldots, 6; w_2 = H, T\}.$$

If the functional form of X_1 and X_2 is

$$\begin{aligned} X_1 &= X_1(w_1) = w_1, & w_1 &= 1, 2, \ldots, 6, \\ X_2 &= X_2(w_2) = 1 & &\text{if } w_2 = H, \\ &= 0 & &\text{if } w_2 = T, \end{aligned}$$

then (X_1, X_2) is a function from Ω onto the rectangle of lattice points

$$\{(a, b) \mid a = 1, 2, \ldots, 6; b = 0, 1\}. \qquad \square$$

Other examples of sample spaces whose points are vectors appeared in Chapter 1 and we now see that such sample spaces arise naturally when considering several random variables simultaneously.

In the important case when both Ω_1 and Ω_2 are the real line, or subsets of the line, the product space $\Omega = \Omega_1 \times \Omega_2$ is the plane or a rectangular subset of the plane. We restrict attention to events (subsets) of Ω which are either rectangles or areas of the plane generated by the formation of countable unions, intersections, and complements of rectangular sets

$$A_{x_1, x_2} = \{(w_1, w_2) \mid w_1 \leqslant x_1, w_2 \leqslant x_2\}, \qquad -\infty < x_1, x_2 < \infty.$$

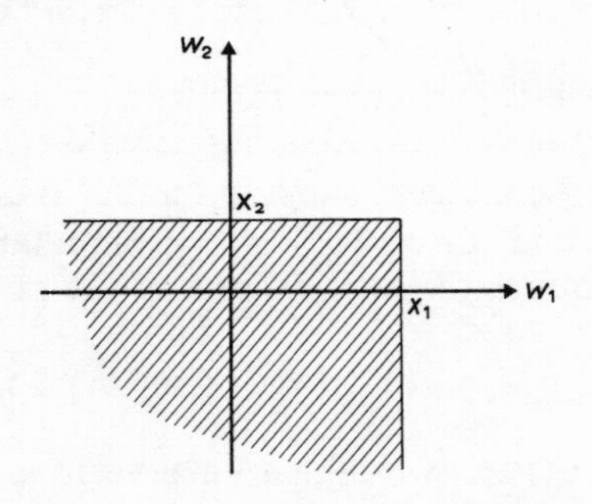

If $\mathscr{F}$ is the σ field containing these events then, as before, we can introduce a probability function P specifying $P(A)$ for all $A \in \mathscr{F}$.

In this case the joint distribution function of $X_1(w_1) = w_1$ and $X_2(w_2) = w_2$ is, from (3.1.3),

$$F_{X_1, X_2}(x_1, x_2) = \Pr(X_1 \leqslant x_1, X_2 \leqslant x_2)$$
$$= P(\{(w_1, w_2) \mid w_1 \leqslant x_1, w_2 \leqslant x_2\}). \qquad (3.1.11)$$

For more general product spaces (3.1.11) must be replaced by

$$F_{X_1, X_2}(x_1, x_2) = P(\{(w_1, w_2) \mid X_1(w_1) \leqslant x_1, X_2(w_2) \leqslant x_2\}). \qquad (3.1.12)$$

The extension of these remarks to the joint consideration of the n random variables $X_1, X_2, \ldots, X_n$, is apparent. If $X_j = X_j(w_j)$ is defined on $\Omega_j = \{w_j\}, j = 1, 2, \ldots, n$, then $(X_1, X_2, \ldots, X_n)$ is a vector function from the product space.

$$\Omega = \Omega_1 \times \Omega_2 \times \ldots \times \Omega_n = \{(w_1, w_2, \ldots, w_n) \mid w_j \in \Omega_j, j = 1, 2, \ldots, n\}$$

into n-dimensional Euclidean space. The joint distribution function is given by (3.1.10) with $\omega = (w_1, w_2, \ldots, w_n)$.

Exercises

1. (Continuation of Example 1.) Assuming that the dice are fair find the probabilities that

(a) $X = 5, Y = 1$, (b) $X = 7, Y \geqslant 3$,
(c) $X = Y$, (d) $X + Y = 4, \quad X - Y = 2$.

2. The same marginal distributions can be obtained from different joint distributions. Hence it is impossible to reconstruct joint probabilities from marginal probabilities alone. Verify that if $f_X(x)$, $f_Y(y)$ are density functions and $g(x, y)$ is any real function of x and y such that

$$|g(x, y| \leqslant f_X(x)\, f_Y(y)$$

$$\int_{-\infty}^{\infty} \int_{-\infty}^{\infty} g(x, y)\, dx dy = 0,$$

then

$$f(x, y) = f_X(x) f_Y(y) + g(x, y)$$

is a bivariate density. In particular show that if a is any constant $|a| < 1$,

$$f(x, y) = 1 + a(1 - 2x)(1 - 2y), \qquad 0 < x, y < 1,$$

is a bivariate density with equal marginal densities. Find the correlation coefficient ρ.
Ref: PARZEN, pages 279, 292.

3. An urn contains 3 white, 1 black, and 2 red balls. A sample of 2 is drawn without replacement. If X and Y are respectively the indicator functions of drawing a white and a black ball, obtain their joint and marginal distributions by completing the following two way table for $\Pr(X = x, Y = y)$

	$X = 0$	$X = 1$	$\Pr(Y = y)$
$Y = 0$			
$Y = 1$			
$\Pr(X = x)$			

4. Suppose $\Pr(X = x) = \frac{1}{4}, x = \pm 1, \pm 2$. If $Y = X^2$ write down the joint distribution of X and Y. Show that their correlation coefficient $\rho_{X,Y}$ is zero even though the variables are functionally related. Thus 'correlated' as defined here must be distinguished from the everyday meaning of the term.

5. (Continuation of Example 2.) Verify

(a) $E(XY) = \rho$
(b) $E \exp\{-(s_1 X + s_2 Y)\} = \exp\{\frac{1}{2}(s_1^2 + 2\rho s_1 s_2 + s_2^2)\}$

6. Show that Schwarz's Inequality implies $(E(XY))^2 \leqslant (EX^2)(EY^2)$ and hence $|\rho_{X,Y}| \leqslant 1$.

7. Show that

$$\begin{aligned}
&\text{(a) } E(X+Y) &&= EX + EY \\
&\text{(b) } \operatorname{Var}(X+Y) &&= \operatorname{Var} X + \operatorname{Var} Y + 2\operatorname{Cov}(X, Y) \\
&\text{(c) } \operatorname{Var}(X-Y) &&= \operatorname{Var} X + \operatorname{Var} Y - 2\operatorname{Cov}(X, Y).
\end{aligned}$$

Observe that expectation always sums whereas this is true for variances only when $\operatorname{Cov}(X, Y) = 0$. Extend to the case of n variables.

8. Show that if c is chosen appropriately then

$$f(x, y) = c\exp[-\tfrac{1}{3}\{6(x+1)^2-2(x+1)(y-2)+(y-2)^2\}],$$

$$-\infty < x, y < \infty$$

is a bivariate normal density function. Find the means, variances, and correlation coefficient.

9. If $\Omega = \Omega_1 \times \Omega_2 = \{(w_1, w_2) \mid w_1 \in \Omega_1, w_2 \in \Omega_2\}$, use (3.1.12) to obtain expressions for the following in terms of the sets

$$A_{x_1, x_2} = \{(w_1, w_2) \mid X_1 \leqslant x_1, X_2 \leqslant x_2\},$$

(a) $\Pr(a_1 < X_1 \leqslant b_1, a_2 < X_2 \leqslant b_2)$
(b) $\Pr(X_1 > 0, X_2 \leqslant x_2)$
(c) $\Pr(X_1 > a_1, X_2 > b_1)$.

10. Show that $f(x, y) = e^{-y}$, $0 < x < y < \infty$, is a bivariate density function. Find the marginal densities and the joint moment generating function.

3.2 CONDITIONAL PROBABILITY

Return to the notation used at the beginning of the previous section and let X and Y be two random variables defined on $(\Omega, \mathscr{F}, P)$. Suppose for the time being that Ω is discrete. Then, using (3.1.2),

$$\Pr(X = x, Y = y) = P(\{\omega \mid X = x, Y = y\})$$

$$= P(\{\omega \mid X = x\}) \frac{P(\{\omega \mid X = x, Y = y\})}{P(\{\omega \mid X = x\})}$$

$$= P(\{\omega \mid Y = y\}) \frac{P(\{\omega \mid X = x, Y = y\})}{P(\{\omega \mid Y = y\})}.$$

Since $P(\{\omega \mid X = x, Y = y\}) = P(\{\omega \mid X = x\} \cap \{\omega \mid Y = y\})$, the quotients on the right hand sides will be recognised as the ratio of the probability that both of two events occur to the probability that a particular one occurs. Such quotients play an important part in the theory and are called *conditional probabilities*.

More formally

$$\frac{\Pr(X = x, Y = y)}{\Pr(X = x)} = \frac{P(\{\omega \mid X = x\} \cap \{\omega \mid Y = y\})}{P(\{\omega \mid X = x\})}$$

is said to be the *conditional probability* that the random variable Y realises the value y, given that $X = x$. It is denoted by the symbol $\Pr(Y = y \mid X = x)$. In the same way the conditional probability that $X = x$, given $Y = y$, is

$$\Pr(X = x \mid Y = y) = \frac{\Pr(X = x, Y = y)}{\Pr(Y = y)}$$

$$= \frac{P(\{\omega \mid X = x\} \cap \{\omega \mid Y = y\})}{P(\{\omega \mid Y = y\})}. \quad (3.2.1)$$

The conditional probability $\Pr(X = x \mid Y = y)$, is not defined if the 'conditioning event' $\{\omega \mid Y = y\}$ has zero probability.

As x varies over the range of X we have for each fixed y that $\Pr(X = x \mid Y = y)$ is a probability distribution, since by the definition of a marginal distribution ((3.1.4)).

$$\sum_x \Pr(X = x \mid Y = y) = \frac{\sum_x \Pr(X = x, Y = y)}{\Pr(Y = y)}$$

$$= \frac{\Pr(Y = y)}{\Pr(Y = y)} = 1.$$

Thus $\{\Pr(X = x \mid Y = y)\}$ is an ordinary function of the discrete variable y and a probability distribution in x. *The conditional distribution function of* X is $\Pr(X \leqslant x \mid Y = y) = \sum_{j \leqslant x} \Pr(X = j \mid Y = y)$.

The conditional probability $\Pr(X = x \mid Y = y)$ defined in (3.2.1) can be restated in terms of events. Writing $A = \{\omega \mid X = x\}$ and $B = \{\omega \mid Y = y\}$, (3.2.1) is

$$P(A \mid B) = P(AB)/P(B),$$

which defines the conditional probability of the event A, given the event B, i.e. the probability that A occurs, given that B has occurred. Clearly there is no need to restrict ourselves to discrete sample spaces and we have

Definition 3.2.1 If $(\Omega, \mathscr{F}, P)$ is a probability space and A and B are events in $\mathscr{F}$, the conditional probability of A, given B, is

$$P(A \mid B) = P(AB)/P(B). \tag{3.2.2}$$

If $P(B) = 0$ then $P(A \mid B)$ is not defined. □

Example 1 Take $B = \Omega$. Then

$$P(A \mid \Omega) = P(A\Omega)/P(\Omega) = P(A)$$

since $A\Omega = A$ and $P(\Omega) = 1$. Thus all probabilities discussed so far can be interpreted as conditional probabilities, given the sample space. In this light (3.2.2) means, as it were that conditioning by an event B shrinks the sample space from Ω to B and conditional probabilities are the probabilities appropriate to this restricted sample space. □

Example 2 Finite sample space with equally likely sample points (§1.3). In this case recall (1.3.3) that the probability $P(A)$ of an event A is given by

$$P(A) = m_A/m,$$

where m_A is the number of sample points in A and m is the total number of sample points. Then

$$P(A \mid B) = \frac{P(AB)}{P(B)}$$

$$= \frac{m_{AB}/m}{m_B/m} = \frac{m_{AB}}{m_B}.$$ □

Example 3 The two-way representation used below is sometimes helpful. Suppose (as in Example 3.1.3) that X can take the values $1, 2, 3, 4, 5, 6$, and Y the values $0, 1$. Suppose also that all twelve sample points are equally likely i.e. $\{\Pr(X = x, Y = y)\}$ is a bivariate uniform distribution over the set $\{(x, y) \mid x = 1, 2, \ldots, 6; y = 0, 1.\}$.

						Row unions
(1.1)	(2.1)	(3.1)	(4, 1)	(5, 1)	(6, 1)	$\bigcup_{x=1}^{6}\{(x, 1$
(1, 0)	(2, 0)	(3, 0)	(4, 0)	(5, 0)	(6, 0)	$\bigcup_{x=1}^{6}\{(x, 0$
Column unions						
$\bigcup_{y=0}^{1}\{(1, y)\}$	$\bigcup_{y=0}^{1}\{(2, y)\}$	$\bigcup_{y=0}^{1}\{(3, y)\}$	$\bigcup_{y=0}^{1}\{(4, y)\}$	$\bigcup_{y=0}^{1}\{(5, y)\}$	$\bigcup_{y=0}^{1}\{(6, y)\}$	Ω

The marginal distribution of Y assigns probabilities to the two events given by the row unions, and similarly the marginal distribution of X is defined on the column unions. Conditional probabilities are obtained by considering only sample points lying in the appropriate row or column. For example $\Pr(X = 3 | Y = 1) = \frac{1}{6}$ since all six points in the upper row are equally likely. On the other hand, using (3.2.2),

$$\Pr(X = 3 \mid Y = 1) = \frac{\Pr(X = 3, Y = 1)}{\Pr(Y = 1)}$$

$$= \frac{P(\{(3, 1)\})}{P\left(\bigcup_{x=1}^{6}\{(x, 1)\}\right)}$$

$$= \frac{1/12}{1/2} = \frac{1}{6}$$

as before. As another example consider $\Pr(X = 2 | X + Y = 3)$. Coordinates of (x, y) have constant sums along diagonals of the table. Since all points are equally likely and precisely two points constitute a diagonal with sum 3, the required probability is $\frac{1}{2}$. □

Returning to Definition 3.2.1, note that it places no restriction on the sample space. Suppose then that X and Y are two random variables on $(\Omega, \mathscr{F}, P)$ and that C_x, C_y are two subsets of the real line such that $A_x = \{\omega \mid X(\omega) \in C_x\}$ and $B_y = \{\omega \mid Y(\omega) \in C_y\}$ are both in $\mathscr{F}$. Substituting $A_x = A$ and $B_y = B$ in (3.2.2) yields

$$P(A_x \mid B_y) = \frac{P(A_x B_y)}{P(B_y)},$$

which in terms of the random variables X and Y is

$$\Pr(X \in C_x \mid Y \in C_y) = \frac{\Pr(X \in C_x, Y \in C_y)}{\Pr(Y \in C_y)}. \tag{3.2.3}$$

This equation is a straightforward generalisation of (3.2.1) and reduces to it if Ω is discrete and $C_x = x$, $C_y = y$. If X and Y are continuous with joint density $f(x, y)$ the choice $C_x = (x, x+dx]$, $C_y = (y, y+dy]$ gives

$$\begin{aligned}\Pr(x < X \leqslant x+dx \mid y < Y \leqslant y+dy) \\ &= \frac{\Pr(x < X \leqslant x+dx, y < Y \leqslant y+dy)}{\Pr(y < Y \leqslant y+dy)} \\ &= \frac{f(x, y)\,dxdy+o(dxdy)}{f_Y(y)\,dy+o(dy)}\end{aligned}$$

where $f_Y(y)$ is the marginal density of Y (assumed to be non-zero at y). To the first order of magnitude

$$\Pr(x < X \leqslant x+dx \mid y < Y \leqslant y+dy) = \frac{f(x, y)}{f_Y(y)}dx,$$

and it has become a generally accepted convention to write the left hand side as

$$\begin{aligned}\Pr(x < X \leqslant x+dx \mid y < Y \leqslant y+dy) \\ &= \Pr(x < X \leqslant x+dx \mid Y = y).\end{aligned}$$

Introducing the notation

$$\Pr(x < X \leqslant x+dx \mid Y = y) = f_{X|Y}(x \mid y)\,dx+o(dx)$$

we have

$$f_{X|Y}(x \mid y)\,dx = \frac{f(x, y)}{f_Y(y)}dx. \tag{3.2.4}$$

Definition 3.2.2 If X and Y are continuous random variables and the marginal density $f_Y(y)$ is positive at y, the function $f_{X|Y}(x|y)$ in (3.2.4) is the *conditional density function of* X, *given* $Y = y$. The *conditional distribution function of* X, *given* $Y = y$, is

$$\Pr(X \leqslant x \,|\, Y = y) = F_{X|Y}(x|y) = \int_{-\infty}^{x} f_{X|Y}(u|y)\,du. \quad (3.2.5)\square$$

Comparing (3.2.1) and (3.2.4) we see that in both the discrete and continuous cases the joint probability (or joint density) can be written as a product of a marginal and a conditional probability (or density),

$$\begin{aligned} \Pr(X = x, Y = y) &= \Pr(Y = y)\Pr(x = x \,|\, Y = y) \\ &= \Pr(x = x)\Pr(Y = y \,|\, X = x) \\ f(x, y) &= f_Y(y)\, f_{X|Y}(x|y) = f_X(x)\, f_{Y|X}(y|x). \end{aligned} \quad (3.2.6)$$

The fundamental concept of independence arises when the right hand sides of (3.2.6) reduce to the product of the two marginal probabilities i.e. when

$$\begin{aligned} \Pr(X = x, Y = y) &= \Pr(X = x)\Pr(Y = y), \\ f(x, y) &= f_X(x) f_Y(y). \end{aligned} \quad (3.2.7)$$

Definition 3.2.3 The random variables X and Y on $(\Omega, \mathscr{F}, P)$ are *independent* if and only if (3.2.7) holds. If they are not independent they are said to be *dependent*. In the same way the events A and B said to be independent or dependent according as $P(AB) = P(A)P(B)$ or not.*
$\square$

This defines the term independence as used in probability theory. To reconcile it as far as possible with everyday usage consider the meaning one may attach to the conditional probability $\Pr(X = x \,|\, Y = y)$ that X takes the value x, given that Y takes the the value y. Intuitively, if X and Y are related in the sense that the value assumed by Y influences that realised by X, one expects that

* The conditional probability $P(A|B)$ is defined only when $P(B) > 0$, in which case $P(A|B) = P(A)$ if the events are independent. However the definition of independence remains valid whether or not $P(B) = 0$.

the probability of X realising a particular value will depend on the prior knowledge that $Y = y$. On the other hand if knowledge of Y is irrelevant to the behaviour of X then it seems plausible that the probability of X realising a particular value should not depend on realisations of Y. This 'irrelevance' of Y to X is expressed mathematically by $\Pr(X = x \mid Y = y) = \Pr(X = x)$ and one arrives at (3.2.7).

Example 4 Let the joint density of X and Y be

$$f(x, y) = \frac{e^{-\frac{1}{2}(x^2+y^2)}}{2\pi}, \qquad -\infty < x, y < \infty.$$

The marginal densities are

$$f_X(x) = \frac{e^{-\frac{1}{2}x^2}}{(2\pi)^{\frac{1}{2}}}, \qquad f_Y(y) = \frac{e^{-\frac{1}{2}y^2}}{(2\pi)^{\frac{1}{2}}}.$$

X and Y are therefore independent since

$$f(x, y) = f_X(x) f_Y(y). \qquad \square$$

Example 5

$$f(x, y) = \frac{1}{2\pi(1-\rho^2)^{\frac{1}{2}}} \exp\left[-\frac{(x^2 - 2\rho xy + y^2)}{2(1-\rho^2)}\right], \qquad -\infty < x, y < \infty.$$

From Example 3.1.2 the marginal densities are

$$f_X(x) = \frac{e^{-\frac{1}{2}x^2}}{(2\pi)^{\frac{1}{2}}}, \qquad f_Y(y) = \frac{e^{-\frac{1}{2}y^2}}{(2\pi)^{\frac{1}{2}}}.$$

But

$$\begin{aligned} f_{X|Y}(x \mid y) &= \frac{f(x, y)}{f_Y(y)} \\ &= \frac{1}{(2\pi(1-\rho^2))^{\frac{1}{2}}} \exp\left[-\frac{(x-\rho y)^2}{2(1-\rho^2)}\right] \\ &\neq f_X(x). \end{aligned}$$

X and Y are therefore dependent, and the conditional distribution of X given $Y = y$ is $N(\rho y, 1-\rho^2)$. $\square$

Example 6 Suppose the random variables X_1 and X_2 are defined on the product space

$$\Omega = \Omega_1 \times \Omega_2 = \{(w_1, w_2) \mid w_1 \in \Omega_1, w_2 \in \Omega_2\}$$

as in the latter part of § 3.1. Then the vector random variable (X_1, X_2) is defined on the probability space $(\Omega, \mathscr{F}, P)$. Recall that $X_1 = X_1(w_1)$ varies only with the first coordinate of $\omega = (w_1, w_2)$, and $X_2 = X_2(w_2)$ only with the second. If C_1 and C_2 are subsets of the real line such that the event $\{(w_1, w_2) \mid X_1(w_1) \in C_1, X_2(w_2) \in C_2\} \in \mathscr{F}$ then

$$\Pr(X_1 \in C_1, X_2 \in C_2) = P(\{(w_1, w_2) \mid X_1 \in C_1, X_2 \in C_2\})$$
$$= P(\{w_1, w_2) \mid X_1 \in C_1, |X_2| < \infty\} \cap \{(w_1, w_2) \mid |X_1| < \infty, X_2 \in C\}) \tag{3.2.8}$$

We recognise the marginal probabilities

$$\Pr(X_1 \in C_1) = P(\{(w_1, w_2) \mid X_1 \in C_1, |X_2| < \infty\})$$

$$\Pr(X_2 \in C_2) = P(\{(w_1, w_2) \mid |X_1| < \infty, X_2 \in C_2\}).$$

If the events in the right-hand side of (3.2.8) are independent (i.e. if X_1 and X_2 are independent) then

$$\Pr(X_1 \in C_1, X_2 \in C_2) = \Pr(X_1 \in C_1)\Pr(X_2 \in C_2),$$

and conversely. We therefore have the result that random variables defined on a product space Ω are independent if and only if the probability function P also factors, $P = P_1 \times P_2$, with P_1 assigning probabilities to $X_1(w_1)$ only and P_2 to $X_2(w_2)$ only. □

Exercises

1. If A, B, C are three events on $(\Omega, \mathscr{F}, P)$ verify that the general results (1.2.1)–(1.2.5) hold also for conditional probabilities, that is, show that

$$P(A \cup B \mid C) = P(A \mid C) + P(B \mid C) - P(AB \mid C)$$

etc.

2. Establish
   (a) $P(ABC) = P(A)\, P(B \mid A)\, P(C \mid AB)$,
   (b) $P(A) = \sum_j P(A \mid B_j)\, P(B)_j)$, if the B_j are disjoint and $\bigcup_j B_j = \Omega$.

   Deduce that if $Y, X_1, X_2, \ldots$ are random variables then

$$\Pr(Y = y) = \sum_j \Pr(Y = y \mid X_j = x_j) \Pr(X_j = x_j)$$

or, in the continuous case,

$$f_Y(y) = \int_{-\infty}^{\infty} f_{Y|X}(y \mid x) f_X(x)\, dx.$$

3. Extend 2(a) above to find

$$P(A_1 A_2 \ldots A_n)$$
$$= P(A_1)P(A_2 \mid A_1)P(A_3 \mid A_1 A_2) \ldots P(A_n \mid A_1 A_2 \ldots A_{n-1}).$$

Show how this expression can be used to interpret sampling without replacement (§ 1.3) in terms of conditional probability.

4. Two dice are rolled. What is the probability that

(a) one face is a six, given that the sum of the two faces is nine,
(b) both numbers shown are odd, given the sum is even,
(c) a six is shown, given both faces are different?

5. A coin is tossed until the first head appears. Given that the first two tosses resulted in tails, what is the probability that more than four tosses are necessary?

6. In a bridge game North's hand includes one ace. What is the probability that his partner (a) has the other three aces, (b) has no ace?

7. Independent random variables are uncorrelated. Show that the converse is false by considering $Y_1 = \sin X$, $Y_2 = \cos X$, where X is uniformly distributed on $(0, 2\pi)$.

8. If $Y_1 = a_{11}X_1 + a_{12}X_2$, $Y_2 = a_{21}X_1 + a_{22}X_2$, where X_1, X_2 are dependent variables, find conditions to ensure that Y_1 and Y_2 are uncorrelated.

9. X is a positive random variable with distribution function $F(x)$ and density $f(x)$. Show that to the first order of magnitude

$$\mu(x)\,dx = \frac{f(x)\,dx}{1-F(x)} = \Pr(x < X \leqslant x+dx \mid x < X).$$

In actuarial circles $\mu(x)$ is called the *force of mortality*. Obtain the formula

$$F(x) = 1-\exp\left(-\int_0^x \mu(y)\,dy\right)$$

and deduce that the negative exponential is the only absolutely continuous distribution function with constant force of mortality.

10. Let $E(X \mid y) = \int_{-\infty}^{\infty} x\, f_{X \mid Y}(x \mid y)\,dx$ be the conditional expectation of X, given $Y = y$. Show that $EX = \int_{-\infty}^{\infty} E(X \mid y) f_Y(y)\,dy$. Denoting expectation with respect to Y by the symbol E_Y, this can be written as $EX = E_Y\{E(X \mid Y)\}$.

11. (Continuation.) In the same way

$$\operatorname{Var} X = E_Y\{\operatorname{Var}(X \mid Y)\} + \operatorname{Var}\{E(X \mid Y)\}.$$

12. (Generalisation of Example 5.) If X and Y have respectively means μ_X, μ_Y; variances σ_X^2, σ_Y^2; and correlation coefficient ρ, they are said to be jointly normally distributed if their joint density is

$$f(x,y) = \frac{1}{2\pi\sigma_X\sigma_Y(1-\rho^2)^{\frac{1}{2}}}$$

$$\exp\left[-\frac{1}{2(1-\rho^2)},\left\{\left(\frac{x-\mu_X}{\sigma_X}\right)^2 - 2\rho\left(\frac{x-\mu_X}{\sigma_Y}\right)\left(\frac{y-\mu_Y}{\sigma_Y}\right) + \left(\frac{y-\mu_Y}{\sigma_Y}\right)^2\right\}\right],$$

$$-\infty < x, y < \infty.$$

Find the conditional density of X, given $Y = y$, and verify that the conditional expectation of X is

$$E(X \mid Y = y) = \mu_X + \rho\sigma_X(y-\mu_Y)/\sigma_Y$$

and the conditional variance

$$\operatorname{Var}(X \mid Y = y) = \sigma_X^2(1-\rho^2).$$

In the statistical literature $E(X \mid Y = y)$ is called the *regression of X on Y* and is used to predict values of X on the basis of realisations of Y. Note that in the normal case the regression is a linear function of y.

13. Let the joint density of X and Y be $f(x, y)$ and *assume* that the regression of X and Y is linear, i.e. $E(X \mid Y = y) = \alpha + \beta y$. Least squares estimation of α and β implies choosing those α and β, say $\alpha = \hat{\alpha}$, $\beta = \hat{\beta}$, that minimise

$$\int_{-\infty}^{\infty}\int_{-\infty}^{\infty} \{E(X \mid Y = y) - (\alpha + \beta y)\}^2 f(x, y)\, dx dy.$$

Show that the least squares procedure leads to the same expression as in Exercise 12, namely

$$\hat{\alpha} = \mu_X - \frac{\rho\sigma_X\mu_Y}{\sigma_Y}, \qquad \hat{\beta} = \frac{\rho\sigma_X}{\sigma_Y},$$

where ρ is the correlation between X and Y, and μ_X, μ_Y their means, σ_X^2, σ_Y^2 their variances, respectively. If $\rho = 0$ the regression line is inappropriate for prediction.

14. For jointly distributed X, Y show that

$$\mathrm{Cov}(X, Y) = \int (y - EY)\, E(X \mid Y = y) f_Y(y)\, dy.$$

Conclude that if $E(X \mid Y = y)$ is independent of y then $\mathrm{Cov}(X, Y) = 0$, and if it is linear, then the covariance is proportional to the variance of Y.

15. The number of calls $X(t)$ at a telephone exchange in a time period of length t is Poisson with parameter λt. Assuming that the number of calls in disjoint time intervals are independent find

(a) the distribution of $X(t_1 + t_2)$

(b) the conditional distribution of $X(t_1)$ given $X(t_1 + t_2) = k$.

16. The conditional expectation $Z = E(X \mid Y)$ is itself a random

variable. For X, Y distributed as in Exercise 12, find the mean, variance, and moment generating function of Z.

3.3 INDEPENDENT RANDOM VARIABLES

We will refer to the notion of independence contained in Definition 3.2.3 as *pairwise independence* since it concerned only two random variables X and Y. If $X_1, X_2, \ldots, X_n$, are randon variables on $(\Omega, \mathscr{F}, P)$ and if $C_1, C_2, \ldots, C_n$ are any subsets of the real line such that

$$\{\omega \mid X_1(\omega) \in C_1, X_2(\omega) \in C_2, \ldots, X_n(\omega) \in C_n\}$$

is in $\mathscr{F}$, then Definition 3.2.3 extends as follows:

Definition 3.3.1 $X_1, X_2, \ldots, X_n$ are independent random variables if and only if

$$P(\{\omega \mid X_j(\omega) \in C_j, j = 1, 2, \ldots, n\}) = \prod_{j=1}^{n} P(\{\omega \mid X_j(\omega) \in C_j\}),$$

or equivalently,

$$\Pr(X_j \in C_j, j = 1, 2, \ldots, n) = \prod_{j=1}^{n} \Pr(X_j \in C_j). \qquad \square$$

Put another way, for the mutual independence of the n random variables we must have the factorisation

$$\begin{aligned} P(\{\omega \mid X_j(\omega) \in C_j, j = 1, 2, \ldots, n\}) &= P\Big(\bigcap_{j=1}^{n} \{\omega \mid X_j(\omega) \in C_j\}\Big) \\ &= \prod_{j=1}^{n} P(\{\omega \mid X_j(\omega) \in C_j\}). \end{aligned}$$

Note that if $X_1, X_2, \ldots, X_n$ are independent then any subset $X_{i_1}, X_{i_2}, \ldots, X_{i_k}$, $k < n$, also has this property. To verify this put $C_j = (-\infty, \infty)$ for $j \neq i_1, i_2, \ldots, i_k$. Then from the definition of a marginal distribution and Definition 3.3.1,

$$\Pr(X_{i_r} \in C_{i_r}, r = 1, 2, \ldots, k) = \prod_{r=1}^{k} \Pr(X_{i_r} \in C_{i_r}).$$

In particular, the mutual independence of $X_1, X_2, \ldots, X_n$ implies the pairwise independence (as in Definition 3.2.3) of all X_j, X_k

$j, k = 1, 2, \ldots, n; j \neq k$. However, the converse is not true. That is, pairwise independence does not imply mutual independence, as the following example shows.

Example 1 $\Omega = \{\omega_1, \omega_2, \omega_3, \omega_4\}$ and each sample point is equally likely. Define the random variables X_1, X_2, X_3 as indicator functions

$$\begin{aligned} X_1(\omega) &= 1 \text{ if } \omega \in \{\omega_1, \omega_2\}, \\ &= 0 \text{ otherwise}, \\ X_2(\omega) &= 1 \text{ if } \omega \in \{\omega_1, \omega_3\}, \\ &= 0 \text{ otherwise}, \\ X_3(\omega) &= 1 \text{ if } \omega \in \{\omega_1, \omega_4\}, \\ &= 0 \text{ otherwise}, \end{aligned}$$

with distributions

$$\begin{aligned} \Pr(X_1 = 1) = \tfrac{1}{2} = \Pr(X_2 = 1) = \Pr(X_3 = 1) \\ \Pr(X_1 = 0) = \tfrac{1}{2} = \Pr(X_2 = 0) = \Pr(X_3 = 0). \end{aligned}$$

Then

$$\begin{aligned} \Pr(X_1 = 1, X_2 = 1) &= P(\{\omega_1\}) = \tfrac{1}{4} \\ &= \Pr(X_1 = 1, X_3 = 1) = \Pr(X_2 = 1, X_3 = 1). \end{aligned}$$

Indeed, since $\frac{1}{4} = \frac{1}{2} \cdot \frac{1}{2}$, it is easily verified that the three random variables are pairwise independent with

$$\Pr(X_1 = j, X_2 = k) = \Pr(X_1 = j)\Pr(X_2 = k) = \tfrac{1}{4}, \quad j, k = 0, 1.$$

On the other hand

$$\begin{aligned} \Pr(X_1 = 1, X_2 = 1, X_3 = 1) &= P(\{\omega \mid X_1 = 1, X_2 = 1, X_3 = 1\}) \\ &= P(\{\omega_1\}) = \tfrac{1}{4}, \end{aligned}$$

but

$$\Pr(X_1 = 1)\Pr(X_2 = 1)\Pr(X_3 = 1) = \tfrac{1}{8}.$$

Hence pairwise independence does not imply the mutual independence of all three random variables. In terms of conditional probabilities we have, for example,

$$\begin{aligned} \Pr(X_1 = 1 \mid X_2 = 1) &= \frac{\Pr(X_1 = 1, X_2 = 1)}{\Pr(X_2 = 1)} \\ &= \frac{P(\{\omega_1\})}{P(\{\omega_1\} \cup \{\omega_3\})} \\ &= \tfrac{1}{4} / \tfrac{1}{2} = \Pr(X_1 = 1) \end{aligned}$$

as it should, but

$$\Pr(X_1 = 1 \mid X_2 = 1, X_3 = 1) = \frac{\Pr(X_1 = 1, X_2 = 1, X_3 = 1)}{\Pr(X_2 = 1, X_3 = 1)} = \frac{P(\{\omega_1\})}{P(\{\omega_1\})} = 1.$$

Put colloquially, the knowledge that both X_2 and X_3 have taken the value 1 yields a different probability for $X_1 = 1$ to that implied by knowing that only X_2 realised the value 1. □

Certain cases of Definition 3.3.1 deserve specific mention. If $C_j = (-\infty, x_j], j = 1, 2, \ldots, n,$ then independence is equivalent to the factorisation of the joint distribution function

$$\Pr(X_1 \leqslant x_1, X_2 \leqslant x_2, \ldots, X_n \leqslant x_n) = \prod_{j=1}^{n} \Pr(X_j \leqslant x_j) \quad (3.3.1)$$

By the same token, in the discrete and continuous cases respectively, independence is necessary and sufficient for

$$\Pr(X_1 = x_1, X_2 = x_2, \ldots, X_n = x_n) = \prod_{j=1}^{n} \Pr(X_j = x_j),$$
$$f(x_1, x_2, \ldots, x_n) = \prod_{j=1}^{n} f_{X_j}(x_j). \quad (3.3.2)$$

In terms of joint generating functions (3.3.2) implies for independent random variables

$$Ez_1^{X_1} z_2^{X_2} \ldots z_n^{X_n} = \prod_{j=1}^{n} \{\sum_{x_j} z_j^{x_j} \Pr(X_j = x_j)\} = \prod_{j=1}^{n} Ez_j^{X_j} \quad (3.3.3)$$
$$E\exp(-(s_1 X_1 + s_2 X_2 + \ldots s_n X_n)) = \prod_{j=1}^{n} E\exp(-s_j X_j).$$

We state the general result in terms of characteristic functions.

Theorem 3.3.1 If $E\exp(i(s_1 X_1 + s_2 X_2 + \ldots + s_n X_n))$ is the joint characteristic function of $X_1, X_2, \ldots, X_n$ then the independence of the n random variables implies

$$Ee^{i(s_1X_1+\ldots+s_nX_n)} = \prod_{j=1}^{n} Ee^{is_jX_j}. \qquad \square$$

Joint moments take on a particularly simple form in the case of independent variables. In particular

$$\frac{\partial^2}{\partial s_j \partial s_k} E\exp(i \sum_{j=1}^{n} s_j X_j)$$

$$= \left[\frac{\partial}{\partial s_j} E\exp(is_j X_j)\right]\left[\frac{\partial}{\partial s_k} E\exp(is_k X_k)\right] \prod_{r=1}^{n} E\exp(is_r X_r)$$

Taking all $s_l = 0$ yields

$$E(X_j X_k) = (EX_j)(EX_k), \tag{3.3.4}$$

implying that the covariance (and correlation) between any pair of independent random variables is zero.* By repeated differentiation and restating the result of Exercise 3.1.7 one has

Theorem 3.3.2 $E(X_1 + X_2 + \ldots + X_n) = EX_1 + EX_2 + \ldots + EX_n$ for any random variables $X_1, X_2, \ldots, X_n$. If $X_1, X_2, \ldots, X_n$ are independent then

$$E(X_1 X_2 \ldots X_n) = (EX_1)(EX_2)\ldots(EX_n) \qquad \square$$

As an application of these results consider the sum of n random variables

$$S_n = X_1 + X_2 + \ldots. + X_n.$$

For each fixed n, S_n is certainly well defined as a random variable on Ω and its distribution function is

$$\Pr(S_n \leqslant x) = P(\{\omega \mid X_1(\omega) + X_2(\omega) + \ldots + X_n(\omega) \leqslant x\})$$

* Hence pairwise independence implies orthogonality. The converse is not true; i.e. $\mathrm{Cov}(X, Y) = 0$ does not in general imply the independence of X and Y (see Exercises 3.2.7). However, in the case of normally distributed variables, independence and orthogonality are equivalent.

If $X_1, X_2, \ldots, X_n$ are independent, does this expression simplify? Firstly note that by Theorem 3.3.1 the characteristic function of S_n is

$$Ee^{isS_n} = Ee^{is(X_1+X_2+\ldots+X_n)} = \prod_{j=1}^{n} Ee^{isX_j}. \tag{3.3.5}$$

Example 2 $X_j \frown N(\mu_j, \sigma_j^2), j = 1, 2, \ldots, n$, and independent.

$$E\exp(isX_j) = \exp\{is\mu_j - \tfrac{1}{2}s^2\sigma_j^2)$$

and

$$E\exp(isS_n) = \exp(is\sum_{j=1}^{n}\mu_j - \tfrac{1}{2}s^2\sum_{j=1}^{n}\sigma_j^2)$$

Hence $S_n \frown N(\sum_{j=1}^{n}\mu_j, \sum_{j=1}^{n}\sigma_j^2)$. □

Whilst a useful result (3.3.5) is not of much help in finding $\Pr(S_n \leqslant x)$ unless, as in the example, one can retrieve the distribution function from the characteristic function. However, comparing (3.3.5) with the results in Chapter 2 on the generating functions of convolutions (Theorems 2.2.4, 2.3.2, and Example 2.3.2) leads one to feel that the distribution of S_n is obtained as the n-fold convolution of the distributions of the summands. We now show that this is the case

Suppose firstly that $n = 2$ and X_1, X_2 are discrete independent variables distributed over the non negative integers. Then

$$\begin{aligned}\Pr(S_2 = k) &= P(\{\omega \mid X_1 + X_2 = k\}) \\ &= P(\bigcup_{j=0}^{k}\{\omega \mid X_1 = j, X_2 = k-j\}) \\ &= \sum_{j=0}^{k} P(\{\omega \mid X_1 = j, X_2 = k-j\}) \\ &= \sum_{j=0}^{k} P(\{\omega \mid X_1 = j\})P(\{\omega \mid X_2 = k-j\}), \\ &\qquad\qquad\text{(by independence)} \\ &= \sum_{j=0}^{k} \Pr(X_1 = j)\Pr(X_2 = k-j).\end{aligned}$$

Hence by Definition 2.2.1 the distribution of S_2 is the convolution of the distributions of X_1 and X_2. Independence is essential here for otherwise we can only infer

$$\Pr(S_2 = k) = \sum_{j=0}^{k} \Pr(X_1 = j, X_2 = k-j).$$

In Exercise 6 you are asked to extend this result and in effect prove the following:

Theorem 3.3.2 If $X_1, X_2, \ldots, X_n$, are independent and S_n is their sum then, if the random variables are discrete,

$$\begin{aligned} \Pr(S_2 = x) &= \sum_y \Pr(X_1 = y)\Pr(X_2 = x-y), \\ \Pr(S_j = x) &= \sum_y \Pr(S_{j-1} = y)\Pr(X_j = x-y), \quad j = 3, 4, \ldots, n. \end{aligned} \tag{3.3.6}$$

In the continuous case the corresponding expression for the density of S_n is given by the iterated convolutions

$$\begin{aligned} f_{S_2}(x) &= \int_{-\infty}^{\infty} f_{X_1}(y) f_{X_2}(x-y)\, dy, \\ f_{S_j}(x) &= \int_{-\infty}^{\infty} f_{S_{j-1}}(y) f_{X_j}(x-y)\, dy, \qquad j = 3, 4, \ldots, n. \end{aligned} \tag{3.3.7}$$ □

To conclude this section we consider the case when Ω can be written as a product space

$$\begin{aligned} \Omega &= \Omega_1 \times \Omega_2 \times \ldots \times \Omega_n \\ &= \{(w_1, w_2, \ldots, w_n) \mid w_j \in \Omega_j, j = 1, 2, \ldots, n\}. \end{aligned}$$

For each $j = 1, 2, \ldots, n$, the random variable $X_j = X_j(w_j)$ depends only on the jth coordinate w_j. Independence of $X_1, X_2, \ldots, X_n$, then implies that the result of Example 3.2.6. generalises and the probability function P can also be factored

$$P = P_1 \times P_2 \times \ldots \times P_n,$$

with each P_j a probability function over Ω_j assigning probabilities to X_j only, $j = 1, 2, \ldots, n$. If in addition all the Ω_j are identical and the P_j are identical, then we may speak of *independent and identically distributed random variables.* For in this case observing the sequence

$X_1, X_2, \ldots, X_n$, is from the probabilistic point of view equivalent to observing n realisations of the same random variable X_1. It is this idea of repeated independent observations of the same random variable that lies at the basis of most statistical reasoning and is expressed in such statements as 'an independent sample from a population', 'repeated independent trials', or 'repetitions of the same experiment'.

If $X_1, X_2, \ldots, X_n$, are independently and identically distributed it will be convenient to say that they are independently and identically distributed as X, where X has distribution function defined by the common value of $P_1, P_2, \ldots, P_n$.

Example 3 Let $S_n = X_1 + X_2 + \ldots + X_n$ where the X_j, $j = 1, 2, \ldots, n$, are independently and identically distributed as X. From (3.3.5) we obtain a formula that will be used frequently in Chapter 4,

$$Ee^{isS_n} = (Ee^{isX})^n. \tag{3.3.8}$$

By differentiation

$$ES_n = nEX, \qquad \operatorname{Var} S_n = n \operatorname{Var} X. \tag{3.3.9}$$ □

Example 4 Coin tossing. Let X be the indicator function

$$\begin{aligned} X &= 1 \quad \text{if heads occur} \\ &= 0 \quad \text{if tails occur} \end{aligned}$$

on the sample space $\{H, T\}$. The statement 'a coin is tossed independently n times and the result of each toss noted' translates into our formalism as a realisation of the random variables $X_1, X_2, \ldots, X_n$. which are independently and identically distributed as X. The sum

$$S_n = X_1 + X_2 + \ldots + X_n$$

denotes the number of heads in these n tosses. If the coin is not necessarily fair with $P(\{H\}) = p$, the distribution of X is

$$\Pr(X = 1) = p, \qquad \Pr(X = 0) = 1 - p$$

with generating function

$$Ez^X = 1 - p - pz$$

From (3.3.8)

$$\begin{aligned} Ez^{S_n} &= (1-p+pz)^n \\ &= \sum_{j=0}^{n} \binom{n}{j} z^j p^j (1-p)^{n-j}. \end{aligned}$$

Hence

$$\Pr(S_n = j) = \binom{n}{j} p^j (1-p)^{n-j}, \qquad j = 0, 1, \ldots, n,$$

and the number of heads obtained in n independent tosses is binomially distributed. □

Independent and identically distributed random variables will be discussed at some length in subsequent chapters. For the time being we return to dependent random variables.

Exercises

1. Show that disjoint events are independent if and only if at least one of them is an event of probability zero.

2. If $A_1, A_2, \ldots, A_n$ are independent events prove that

$$P\left(\bigcup_{j=1}^{n} A_j\right) = 1 - \prod_{j=1}^{n} P(\bar{A}_j).$$

3. Find the distribution functions of

$$\begin{aligned} Y &= \max(X_1, X_2, \ldots, X_n) \\ Z &= \min(X_1, X_2, \ldots, X_n), \end{aligned}$$

where $X_1, X_2, \ldots, X_n$ are independently distributed with common distribution function $F(x)$.

4. Suppose X_1, X_2, and X_3 are independent and identically distributed as X, where

$$\Pr(X = 1) = p, \qquad \Pr(X = 0) = q = 1-p.$$

Write down the probabilities that

(a) exactly one of X_1, X_2, X_3 realises the value one,
(b) only X_1 realises the value one,
(c) at least two ones are realised.

5. (c.f. Example 4) Suppose $P(A_k) = p,\ k = 1, 2, \ldots, n$ and the events $A_1, A_2, \ldots, A_n$ are independent. Show that the probability that exactly j events occur is

$$\binom{n}{j} p^j (1-p)^{n-j}, \qquad j = 0, 1, \ldots, n.$$

6. Use induction to complete the proof of Theorem 3.3.2.

7. A game consists of n independent plays or trials. If the probability of a win at a trial is $\frac{2}{5}$, find the distribution and the mean and variance of the number of wins. Suppose that an entrance fee of \$2 per trial is charged, a win resulting in a prize of \$5. Write down the distribution of the gain over n trials. Do you consider the game fair?

8. You are given that j 'successes' occurred in n independent and identical trials, the outcome of each trial being either 'success' or 'failure'. Show that the conditional probability of success on a specified trial is j/n.

9. An experiment can result in one of three possible outcomes with respective probabilities of occurrence $p_1, p_2, p_3; p_1 + p_2 + p_3 = 1$. In n independent repetitions of the experiment show that the probability of obtaining n_1 outcomes of the first type and n_2 of the second is

$$\frac{n!}{n_1!\, n_2!\, (n - n_1 - n_2)!} p_1^{n_1} p_2^{n_2} p_3^{n-(n_1+n_2)}$$

10. A coin is tossed successively until the first head appears. If the probability of obtaining a head on a single toss is p, show that the distribution of the 'waiting time' for the first head is

$$p_n = (1-p)^{n-1} p, \qquad n = 1, 2, \ldots.$$

[Strictly speaking, in this example we require the sample points to be vectors of infinite dimension. However, for every fixed n the probability that a head does not appear in the first n trials is $(1-p)^n$, which tends to zero as $n \to \infty$. Thus the probability is one that the first head appears at some toss.]

11. (Continuation.) Show that the 'waiting time' until the kth head appears is negative binomially distributed.

12. If $X_1, X_2, \ldots, X_n$ are independent variables identically distributed as X, find the distribution of their sum S_n when

(a) $\Pr(X = x) = \binom{N}{x} p^x(1-p)^{N-x}, \qquad x = 0, 1, 2, \ldots, N,$

(b) $\Pr(X = x) = e^{-\lambda}\dfrac{\lambda^x}{x!} \qquad x = 0, 1, 2, \ldots,$

(c) $\Pr(X = x) = (1-\rho)\rho^x, \qquad x = 0, 1, 2, \ldots,$

(d) X has density $f(x) = \lambda e^{-\lambda x}, \qquad x > 0.$

13. If X_1, X_2 are independent with densities $f_1(x)$, $f_2(x)$ respectively, show that $Y = X_1 - X_2$ has density

$$f(y) = \int_{-\infty}^{\infty} f_1(y+u) f_2(u)\, du.$$

If, also, X_1 and X_2 are identically distributed, show that the density of Y is symmetric about the origin and that its characteristic function is $Ee^{isY} = |Ee^{isX}|^2$ (c.f. Example 2.3.5).

14. (Continuation) Find the density and characteristic function of Y when X_1, X_2 have distributions

(a) $N(\mu, \sigma^2)$,
(b) negative expotential.

15. *Random sums.* Let $S_N = X_1 + X_2 + \ldots + X_N$, where the number of summands is a random variable N on the positive integers, and the X_j are independent and identically distributed as X. Show that $ES_N = (EN)(EX)$, $\operatorname{Var} S_N = (EN)\operatorname{Var} X + (\operatorname{Var} N)(EX)^2$, and find the characteristic function of S_N if N is Poisson. An application in biology interprets N as the number of female insects in a region and X as the number of eggs laid by a female, in which case S_N is the total number of eggs.

3.4* BAYES'S THEOREM

If A and B are two events on $(\Omega, \mathscr{F}, P)$ then from Definition 3.2.1

$$P(AB) = P(A)P(B \mid A) = P(B)P(A \mid B).$$

* To be omitted at a first reading.

Consequently

$$P(A \mid B) = \frac{P(B \mid A)P(A)}{P(B)}. \tag{3.4.1}$$

This equation is known as *Bayes's Theorem* or *Bayes's formula.*†

If $A_1, A_2, \ldots, A_n$ are disjoint and exhaustive (i.e. $\Omega = \bigcup_{j=1}^{n} A_j$) then the events $BA_1, BA_2, \ldots, BA_n$ are also disjoint. In this case

$$\begin{aligned} P(B) &= P(B\Omega) = P\left(\bigcup_{j=1}^{n} BA_j\right) \\ &= \sum_{j=1}^{n} P(BA_j) = \sum_{j=1}^{n} P(B \mid A_j)P(A_j), \end{aligned} \tag{3.4.2}$$

and using (3.4.1) with $A = A_j$,

$$P(A_j \mid B) = \frac{P(B \mid A_j)P(A_j)}{\sum_{j=1}^{n} P(B \mid A_j)P(A_j)}. \tag{3.4.3}$$

Example 1 Three urns contain respectively 1 white, 3 red, and 2 black balls; 3 white, 1 red, and 1 black ball; 3 white, 3 red, and 3 black balls. Two balls are chosen from a randomly selected urn. If the two balls are a white and a red, what is the probability that they came from the second urn?

Let B denote the observed event that the two balls in the sample are a white and a red. Let A_j denote the event that they were selected from urn number j, $j = 1, 2, 3$, and we have to compute the conditional probability $P(A_2 \mid B)$. Firstly, since the urns are randomly chosen,

$$P(A_1) = P(A_2) = P(A_3) = \tfrac{1}{3}.$$

Also

$$P(B \mid A_1) = \tfrac{1}{6}\cdot\tfrac{3}{5}; \quad P(B \mid A_2) = \tfrac{3}{5}\cdot\tfrac{1}{4}; \quad P(B \mid A) = \tfrac{3}{9}\cdot\tfrac{3}{8},$$

and the denominator in (3.4.3) is

$$P(B) = \sum_{j=1}^{3} P(B \mid A_j)P(A_j) = \frac{3}{30}\cdot\frac{1}{3} + \frac{3}{20}\cdot\frac{1}{3} + \frac{9}{72}\cdot\frac{1}{3} = \frac{1}{8}.$$

† After Thomas Bayes (1702–1761). His important paper was published posthumously in 1763.

From (3.4.3) the required solution is

$$P(A_2 \mid B) = \frac{1/20}{1/8} = \frac{2}{5}. \qquad \square$$

Bayes's formula can be used to solve problems such as that of Example 1, but its most interesting application (originally proposed by Bayes) is in the development of a form of inductive inference. This involves a novel interpretation of the probability function P, and we proceed to describe a primitive version of the theory.

Suppose the disjoint and exhaustive hypotheses $\mathscr{A}_1, \mathscr{A}_2, \ldots, \mathscr{A}_n$ concerning a phenomenon of interest are given. For example before tossing a coin we may hypothesise '$\mathscr{A}_1$: the coin is fair' or '$\mathscr{A}_2$: the coin is not fair'. On the evidence available we agree to allocate probabilities $\Pr(\mathscr{A}_j)$ (with $\sum_j \Pr(\mathscr{A}_j) = 1$) to each of these hypotheses as measures of our belief in their validity. That is, not only can we order the n hypotheses according to credibility but we take the further step of assuming that it is possible to measure differences in our beliefs in a numerical way by specifying values for the $\Pr(\mathscr{A}_j), j = 1, 2, \ldots, n$. This point will be taken up again later; for the moment we take such probabilities as given.

An experiment designed to provide more information is carried out and results in the occurrence of what we will call the event B. How should our degree of belief in the hypotheses $\mathscr{A}_1, \mathscr{A}_2, \ldots, \mathscr{A}_n$ be modified due to B occurring?

The answer is given by Bayes's formula (3.4.1) or (3.4.3) provided it is possible to set up a suitable probability space with probability function P. A way of doing this is to construct the sample space Ω as a product

$$\Omega = \Omega_1 \times \Omega_2$$

in which the hypotheses refer to events in Ω_1 and the possible results of the experiment are enumerated in Ω_2. It is convenient to call Ω_2 the *experiment* or *observation space*.

We will call Ω_1 the *parameter space* since in applications the hypotheses $\mathscr{A}_1, \ldots, \mathscr{A}_n$ frequently concern parameters of a distribution function. On Ω_1 we define a probability function P_1 and disjoint and exhaustive events $A_1, A_2, \ldots, A_n$ such that

$$\Pr(\mathscr{A}_j) = P_1(A_j), \qquad j = 1, 2, \ldots, n. \tag{3.4.4}$$

In other words, given the hypotheses $\mathscr{A}_1, \ldots, \mathscr{A}_n$ and the numbers $\Pr(\mathscr{A}_1), \ldots, \Pr(\mathscr{A}_n)$ expressing our degree of belief in their validity, we induce a probability space $(\Omega_1, \mathscr{F}_1, P_1)$ such that (3.4.4) holds.

On the appropriate subsets of the product space $\Omega = \Omega_1 \times \Omega_2$ (= parameter space × observation space) we suppose given a probability function P satisfying the condition that the marginal probabilities of the events $A_j \subset \Omega_1$ are given by

$$P(A_j) = P_1(A_j) = \Pr(\mathscr{A}_j), \qquad j = 1, 2, \ldots, n.$$

These marginal probabilities $P(A_j)$ are called *a priori* or *prior* probabilities since they measure our belief in the various hypotheses before the experiment takes place. For all events $C \subset \Omega_2$ the conditional probabilities $P(A_j \mid C), j = 1, 2, \ldots, n$, are called the *a posteori* or *posterior* probabilities of the corresponding hypotheses, since they represent our reassessment via Bayes's formula of our degree of belief in these hypotheses, given that the experiment resulted in C.

For $C \subset \Omega_2$ the conditional probability $P(C \mid A_j)$ as a function over $\{A_1, A_2, \ldots, A_n\}$ is called the *likelihood of C*. $P(C \mid A_j)$ is the probability of observing C as the outcome of the experiment, given that the hypothesis $\mathscr{A}_j$ is true, i.e. given the event A_j.

The difference (or ratio) of the prior and posterior probabilities of the event A_j measures the change in our belief in the truth of $\mathscr{A}_j$ consequent on carrying out the experiment. In this sense Bayes's formula is a statement in probabilistic terms of how one learns by experience. Clearly it is futile to perform an experiment outcomes of which are independent of the $A_j \subset \Omega_1$, for then the prior probabilities remain unchanged and the experiment is non-informative. On the other hand it may so happen that an experiment can be designed which results in a posterior probability equal to or close to unity for a particular A_k and negligible for the other events in Ω_1. This is frequently the case in experimental science, and the above arguments are not relevant. However there is a large class of phenomena for which decisive experiments cannot be designed, and it is here that inference procedures using Bayes's formula may be helpful.

Example 2 It is known that the probability p of a given coin showing heads when tossed is either equal to p_1 or to p_2. We have the two hypotheses $\mathscr{A}_1$: $p = p_1$ and $\mathscr{A}_2$: $p = p_2$ with parameter space

$\Omega_1 = \{w_1 \mid w_1 = p_1, p_2\}$ and prior probabilities

$$P(A_1) = P_1(\{w_1 \mid w_1 = p_1\}), \quad P(A_2) = P_1(\{w_1 \mid w_1 = p_2\}).$$

To obtain more information the coin is tossed independently n times and it is observed that x of these tosses result in heads. The event B realised by the experiment is then 'x heads in n tosses' and B is contained in the experiment space Ω_2,

$$\Omega_2 = \{w_2 \mid w_2 = 0, 1, 2, \ldots, n\}.$$

The sample space on which the probability function P is defined is

$$\Omega = \Omega_1 \times \Omega_2 = \{(w_1, w_2) \mid w_1 = p_1, p_2 ; w_2 = 0, 1, 2, \ldots, n\}.$$

Recall from Example 3.3.4 that the probability of obtaining x heads in n tosses of a coin is given by the binomial distribution. Then the likelihood of B on the two hypotheses is

$$P(B \mid A_1) = \binom{n}{x} p_1^x (1-p_1)^{n-x}, \quad P(B \mid A_2) = \binom{n}{x} p_2^x (1-p_2)^{n-x},$$

and from Bayes's formula the posterior probabilities are

$$\begin{aligned} P(A_1 \mid B) &= \frac{p_1^x (1-p_1)^{n-x} P(A_1)}{p_1^x (1-p_1)^{n-x} P(A_1) + p_2^x (1-p_2)^{n-x} P(A_2)} \\ P(A_2 \mid B) &= \frac{p_2^x (1-p_2)^{n-x} P(A_2)}{p_1^x (1-p_1)^{n-x} P(A_1) + p_2^x (1-p_2)^{n-x} P(A_2)} \end{aligned} \qquad (3.4.5)$$

Provided the symbols on the right hand sides are specified numerically, it is easy to calculate values for the posterior probabilities. Comparing these with the prior probabilities one discovers which hypotheses are confirmed by the experiment and which are not. □

The example has certain special features which will be followed up in Exercises 4, 5. For the moment consider those aspects more directly relevant to the general theory. Firstly note that the term 'parameter space' for Ω_1 was particularly apt for the example since the hypotheses concerned the unknown parameter p of a binomial distribution. Furthermore p was treated as if it was a random variable on $\{p_1, p_2\}$ with distribution determined by the measures of degree of belief $\Pr(\mathscr{A}_1)$ and $\Pr(\mathscr{A}_2)$.

Two objections immediately come to mind:

(a) A parameter, even if its value is unknown, is nonetheless a constant and should not under any circumstances be treated as a random variable.

(b) How does one measure degree of belief in a hypothesis?

Other criticisms can be raised but these are the obvious ones. It is (b) that is the more serious of the two, and the question has generated some controversy. Those who answer by asserting that in general it is possible to specify the measures of belief $\Pr(\mathscr{A}_j)$ in a rational way are said to hold a subjectivist view in the sense that they admit the validity of personal or subjective probabilities*. Bayes himself assigned prior probability to a hypothesis according to the odds he would lay on the hypothesis being true, and in this sense held a variant of the modern subjectivist theory. The difficulty is frequently to derive a prior distribution from vague information and to develop inference procedures in which this vagueness is taken into account.

Disagreement on probability as degree of belief does not affect our general theory in any way. The difference of opinion concerns interpretation and on how values of a probability function P should be assigned. Whether or not our definition of probability in § 1.1 is sound is not at issue. Questions of interpretation will be briefly taken up again after discussing the laws of large numbers.

Exercises

1. Four balls are removed from an urn containing four white and four red balls and placed in a second urn. Three balls from the second urn are placed in a third urn. A ball is chosen from the third urn and it turns out to be white. What is the probability that all four balls taken from the first urn are white?

2. If X and Y are jointly distributed random variables verify that Bayes's formula can be written as

$$\Pr(X = \,|\, Y = y) = \frac{\Pr(Y = y \,|\, X = x)\Pr(X = x)}{\Pr(Y = y)}$$

* For an introductory account with applications see D. V. LINDLEY *Introduction to Probability and Statistics*, Vol. 1, particularly pages 19–41.

or

$$f_{X|Y}(x \mid y) = \frac{f_{Y|X}(y \mid x) f_X(x)}{f_Y(y)}.$$

3. (Continuation.) Suppose $X \frown N(\mu, \sigma^2)$ where σ^2 is known but the mean μ is regarded as a realisation of a random variable Y. If the prior distribution of Y is $N(\mu_Y, \sigma_Y^2)$ with μ_Y and σ_Y^2 given show that the posterior density of Y is normal with mean

$$(x\sigma^{-2} + \mu_Y \sigma_Y^{-2})/(\sigma^{-2} + \sigma_Y^{-2})$$

and variance $(\sigma^{-2} + \sigma_Y^{-2})^{-1}$.

4. (Continuation of Example 2.) Let the experiment consist of an arbitrarily large number n of tosses. Assuming that the observed number x of heads increases with n to ensure $x \sim \lambda n$, $0 < \lambda < 1$, and that $p_1 + p_2 \neq 1$, show that one of the posterior probabilities approaches the limit 1 and the other approaches 0. Thus there is an experiment which in this asymptotic sense discriminates perfectly between the two hypotheses.

5. We will say that the hypothesis $\mathscr{A}_1$ is confirmed by an experiment resulting in B if $P(A_1 \mid B) > P(A_1)$. Show that in Example 2 $\mathscr{A}_1$ is confirmed by observing x heads in n tosses if and only if

$$p_1^x(1-p_1)^{n-x} > p_2^x(1-p_2)^{n-x}$$

irrespective of the numerical values assigned to the prior probabilities. That is, confirmation of a hypothesis in this example depends only on the likelihood $P(B \mid A_j)$, $j = 1, 2$.

6. Compute the posterior distribution when the likelihood is binomial with parameter p and the prior is beta with parameters a and b. Compare the means and maximum frequencies of the two distributions.

7. *Laplace's law of succession.* Suppose all n tosses of a coin have shown heads. Given this information we require the probability that a further toss will also show heads. If the unknown probability p of obtaining heads at a single toss is assumed to be uniformly

distributed over $\{r/N, r = 1, 2, \ldots, N\}$ show that when N is large the required probability is approximately $(n+1)/(n+2)$.

Ref: FELLER I pages 123–124; PARZEN pages 121–124.]

3.5* SEQUENCES OF DEPENDENT RANDOM VARIABLES; MARKOV CHAINS

Given events $A_0, A_1, A_2, \ldots, A_m$ on $(\Omega, \mathscr{F}, P)$ we know from Exercise 3.2.3 that

$$P(A_0 A_1 \ldots A_m) = P(A_0)P(A_1 \mid A_0)P(A_2 \mid A_0 A_1)\ldots P(A_m \mid A_0 A_1 \ldots A_{m-1}). \tag{3.5.1}$$

If $X_0, X_1, \ldots, X_m$ are discrete random variables such that

$$A_j = \{\omega \mid X_j(\omega) = x_j\}, \qquad j = 0, 1, \ldots, m,$$

(3.5.1) implies that their joint probability is given by the product

$$\begin{aligned}\Pr(X_0 = x_0, X_1 = x_1, \ldots, X_m = x_m)& \\ = \Pr(X_0 = x_0)\Pr(X_1 = x_1 \mid X_0 = x_0)& \\ \ldots \Pr(X_m = x_m \mid X_0 = x_0, \ldots, X_{m-1} = x_{m-1}).& \qquad (3.5.2)\end{aligned}$$

Similarly in the continuous case, with $A_j = \{\omega \mid x_j < X_j \leqslant x_j + dx_j\}$, the joint density can be written

$$\begin{aligned}f(x_0, x_1, \ldots, x_m) = f_{X_0}(x_0) f_{X_1 \mid X_0}(x_1 \mid x_0)& \\ f_{X_m \mid X_0, X_1, X_{m-1}}(x_m \mid x_0, x_1, \ldots, x_{m-1}).& \qquad (3.5.3)\end{aligned}$$

This section treats a specialisation of these formulae. Whilst (3.5.2) and (3.5.3) are generally valid it is convenient for our purposes to take Ω as a product space

$$\begin{aligned}\Omega &= \Omega_0 \times \Omega_1 \times \ldots \times \Omega_m \\ &= \{(w_0, w_1, \ldots, w_m) \mid w_j \in \Omega_j, j = 0, 1, \ldots, m\}\end{aligned}$$

with $X_j = X_j(w_j)$ a function over Ω_j only. It will also be convenient to regard the subscript j as denoting an ordering (usually in time)

* To be omitted at a first reading.

of the variables. In this case X_0 is the first random variable to be realised, X_1 the second, ..., X_m the $(m+1)$th, and we may speak of the (finite) sequence of random variables $\{X_j, j = 0, 1, \ldots, m\}$. Sequences of random variables indexed by time are frequently called *stochastic processes*.

Example 1 A game consists of an independent sequence of plays (or trials) at each of which a gambler can win or lose a unit amount or not bet. A *system* is a set of rules which at every trial determines whether or not the gambler will bet. Clearly any reasonable system takes past (but not future) experience into account. For a given system let $X_j = 1, 0$, or -1 according as the gambler bets and wins, does not bet, or bets and loses, at the jth trial. Then X_j depends only on $X_0, X_1, \ldots, X_{j-1}$ and the history of the game over $m+1$ trials is described in probabilistic terms by (3.5.2)*. □

The most important case of (3.5.2), (3.5.3) as far as current theory is concerned occurs when for $j = 2, 3, \ldots$

$$\Pr(X_j = x_j \,|\, X_0 = x_0, X_1, \ldots, X_{j-1} = x_{j-1}) \\ = \Pr(X_j = x_j \,|\, X_{j-1} = x_{j-1}), \qquad (3.5.4)$$

$$f_{X_j|X_0, X_1 \cdots X_{j-1}}(x_j \,|\, x_0, x_1, \ldots, x_{j-1}) = f_{X_j|X_{j-1}}(x_j \,|\, x_{j-1})$$

Interpreting j as 'time', (3.5.4) asserts that X_j depends only on X_{j-1}, the immediately preceding member, and not on the history of the sequence before time $j-1$. That sequences of random variables with this property arise naturally is shown by the following:

Example 2 Let $Y_1, Y_2, \ldots$ be independent random variables identically distributed as Y. Suppose that Y is distributed over the integers and let the initial value $Y_0 = 0$ be fixed. Put

$$S_n = Y_1 + Y_2 + \ldots + Y_n, \qquad n = 1, 2, \ldots; \; S_0 = 0.$$

We show that the sequence $\{S_n\}$ satisfies (3.5.4); that is,

$$\Pr(S_n = k \,|\, S_{n-1} = i) \\ = \Pr(S_n = k \,|\, S_1 = s_1, S_2 = s_2, \ldots, S_{n-1} = i). \qquad (3.5.5)$$

* A proof that such gambling systems do not improve one's chances of winning is given in FELLER I, page 199.

Firstly note that

$$\begin{aligned}\Pr(S_n = k \mid S_{n-1} = i) &= \frac{\Pr(S_n = k, S_{n-1} = i)}{\Pr(S_{n-1} = i)} \\ &= \frac{\Pr(Y_n = k-i, S_{n-1} = i)}{\Pr(S_{n-1} = i)} \\ &= \Pr(Y_n = k-i) = \Pr(Y = k-i),\end{aligned}$$

the last step following because Y_n is independent of the other Y_j, and hence also independent of S_{n-1}. Further

$$\begin{aligned}\Pr(S_n = k \mid S_{n-1} = i, S_{n-2} = r) &= \frac{\Pr(S_n = k, S_{n-1} = i \mid S_{n-2} = r)}{\Pr(S_{n-1} = i \mid S_{n-2} = r)} \\ &= \frac{\Pr(Y_n = k-i, S_{n-1} = i \mid S_{n-2} = r)}{\Pr(S_{n-1} = i \mid S_{n-2} = r)} \\ &= \Pr(Y_n = k-i) = \Pr(Y = k-i),\end{aligned}$$

again by independence. The argument clearly generalises and we have (3.5.5). A similar argument holds for the sums of independent and identically distributed continuous variables. □

A sequence of random variables for which (3.5.4) holds for all $j = 1, 2, 3, \ldots$ is called a *discrete parameter Markov process or chain**. 'Discrete parameter' refers to the fact that we are taking time to be discrete, counted by the integer valued subscript j. We do not consider continuous time processes. If, also, the conditional probability $\Pr(X_j = k \mid X_{j-1} = i)$ or density $f_{X_j|X_{j-1}}(x \mid y)$ does not depend on the time index j then $\{X_j\}$ is said to be a *homogeneous Markov chain* (e.g. the partial sums $\{S_j\}$ of Example 2).

Let us assume in what follows that $\Omega_0 = \Omega_1 = \ldots = \Omega_m$, in which case Ω is the $m+1$ fold product of Ω_0, and that Ω_0 is discrete. Then all $X_j = X_j(w_j)$ are functions over the same space Ω_0 and we will assume further that they realise values in the same set, say the integers or a proper subset of the integers. This set containing realisations of the random variables is called the *state space* of the process

* After A. A. Markov (1856–1922).

$\{X_j\}$. We restrict attention to these integer valued variables $X_0, X_1, \ldots, X_m$.

Given these conditions and that $\{X_j\}$ is a homogeneous Markov chain, the probabilistic development in time $[0, m]$ of this sequence of random variables is completely known once the values of

$$p_{ik} = \Pr(X_j = k \mid X_{j-1} = i) \tag{3.5.6}$$

and $\Pr(X_0 = x_0)$ are given. For it then follows from (3.5.2) and (3.5.4) that

$$\Pr(X_0 = i, X_1 = x_1, \ldots, X_m = k)$$
$$= \Pr(X_0 = i) p_{ix_1} p_{x_1 x_2} \cdots p_{x_{m-1}k}. \tag{3.5.7}$$

As a function of all i, k in the state space, the conditional probability p_{ik} defined by (3.5.6) is called the (one step) *transition probability function* of the chain $\{X_j\}$.

The important point to note is that it is the initial distribution $\{\Pr(X_0 = x_0)\}$ and the transition probability function $\{p_{ik}\}$ that uniquely determine the probabilistic evolution of a homogeneous Markov chain. Conversely, if we are given an initial distribution $\{p_i\}$ and a function $\{p_{ik}\}$ of the two discrete variables i, k which for every i is a probability distribution (i.e. $0 \leqslant p_{ik} \leqslant 1, \sum_k p_{ik} = 1$), we can induce a Markov chain $\{X_j, j = 0, 1, 2, \ldots\}$ in the same way that probability distributions were used to induce probability spaces and random variables in Chapter 1.

Frequently one is not so much interested in the complete evolution of the process as in the n-step transition probability $p_{ik}^{(n)}$,

$$p_{ik}^{(n)} = \Pr(X_n = k \mid X_0 = i), \tag{3.5.8}$$

which is the probability of the sequence taking the value k at time n, given the initial state i. This is obtained from the joint distribution of X_0 and X_n which is (using (3.5.7))

$$\Pr(X_0 = i, X_n = k) = \sum_{x_1, x_2, \ldots, x_{n-1}} \Pr(X_0 = i, X_1 = x_1, \ldots, X_n = k)$$
$$= \Pr(X_0 = i) \sum_{x_1, x_2, \ldots, x_{n-1}} p_{ix_1} p_{x_1 x_2} \cdots p_{x_{n-1}k}.$$

Hence, from the definition of conditional probability and (3.5.8),

$$p_{ik}^{(n)} = \sum_{x_1, x_2, \ldots, x_{n-1}} p_{ix_1} p_{x_1 x_2} \cdots p_{x_{n-1}k}. \tag{3.5.9}$$

The absolute or unconditional distribution of the sequence at time n is given by

$$\Pr(X_n = k) = \sum_i \Pr(X_0 = i) p_{ik}^{(n)}. \tag{3.5.10}$$

Notation becomes tidier if we introduce the matrix $\mathscr{P}$ of one step transition probabilities

$$\mathscr{P} = (p_{ik}).$$

That is, the element in the ith row and kth column of $\mathscr{P}$ is p_{ik}. $\mathscr{P}$ is a finite matrix if the X_j can realise only a finite number of values (i.e. if the state space is finite); otherwise it is an infinite matrix. Note the important fact that all elements of $\mathscr{P}$ are non-negative (being probabilities) and the sum of the elements in every row is equal to unity. Such matrices are called *stochastic matrices.*

Whether or not $\mathscr{P}$ is finite, matrix multiplication by it is still defined and in particular

$$\mathscr{P} \times \mathscr{P} = \mathscr{P}^2 = (\sum_j p_{ij} p_{jk}).$$

$\mathscr{P}^2$ is again a stochastic matrix since the sum of elements in the ith row is

$$\sum_k \sum_j p_{ij} p_{jk} = \sum_j p_{ij} \sum_k p_{jk} = 1$$

Comparing $\sum_j p_{ij} p_{jk}$ (the (i, k)th element of $\mathscr{P}^2$) with (3.5.9) for $n = 2$, namely

$$p_{ik}^{(2)} = \Pr(X_2 = k | X_0 = i) = \sum_{x_1} p_{ix_1} p_{x_1 k},$$

we conclude that the two are identical. Thus the (i, k)th term of $\mathscr{P}^2$ is equal to $p_{ik}^{(2)}$. The argument generalises and with the n-step transition probability $p_{ik}^{(n)}$ defined in (3.5.8),

$$\mathscr{P}^n = (p_{ik}^{(n)}).$$

Example 3 Let the state space of the Markov chain $\{X_j\}$ be the numbers 0, 1, 2, with transition matrix

$$\mathscr{P} = \begin{pmatrix} q & p & 0 \\ q & 0 & p \\ 0 & q & p \end{pmatrix}, \qquad q+p = 1.$$

Then

$$\mathscr{P}^2 = \begin{pmatrix} q^2+qp & qp & p^2 \\ q^2 & 2qp & p^2 \\ q^2 & qp & qp+p^2 \end{pmatrix}$$

and, for example, $\Pr(X_2 = 0 \mid X_0 = 1) = q^2$. □

Example 4 In Example 2 it was shown that the sequence of partial sums $\{S_n, n = 1, 2, \ldots\}$ of independent variables identically distribution as Y, constituted a homogeneous Markov chain. If the distribution of Y is

$$\Pr(Y = j) = a_j, \qquad j = 0, \pm 1, \pm 2, \ldots,$$

then, from Example 2, the transition probabilities are

$$p_{ik} = \Pr(S_n = k \mid S_{n-1} = i) = \Pr(Y = k-i) = a_{k-i}.$$

In the present context the Markov chain $\{S_n\}$ is called a *random walk* since we can interpret the S_n as successive positions of a particle on the real line subject to random displacements $Y_1, Y_2, \ldots$ at time points $n = 1, 2, \ldots$. The n-step transition probabilities are given by the n-fold convolution of the distribution $\{a_j\}$ since

$$\begin{aligned} p_{ik}^{(n)} &= \Pr(S_n = k \mid S_0 = i) \\ &= \Pr(Y_1 + Y_2 + \ldots + Y_n = k-i). \end{aligned} \qquad (3.5.11)$$

A *restricted random walk* is one in which the sums S_n are constrained to lie in an interval, say $[0, b]$. The transition matrix of such a restricted walk is a $(b+1)\times(b+1)$ stochastic matrix with general term $p_{ik} = a_{k-i}$ but with the extremal rows and columns modified. For example, if

$$a_1 = p, \quad a_{-1} = q = 1-p, \quad a_j = 0 \text{ otherwise},$$

(in which case we call the random walk *simple*) the transition matrix of the unrestricted walk is the doubly infinite matrix

$$\mathscr{P} = \begin{pmatrix} \cdots & & & & & & & & \cdots \\ \cdots & 0 & q & 0 & p & 0\cdots & & & \cdots \\ \cdots & & 0 & q & 0 & p & 0 & & \cdots \\ \cdots & & & 0 & q & 0 & p & 0 & \cdots \\ \cdots & & & & & & & & \cdots \end{pmatrix}$$

On the other hand if the walk is restricted to $[0, b]$ the matrix is

$$\underset{(b+1)\times(b+1)}{\mathscr{P}} = \begin{pmatrix} c_0 & c_1 & c_2 & . & . & . & . & . & c_b \\ q & 0 & p & 0 & . & . & . & . & 0 \\ 0 & q & 0 & p & 0 & & & & 0 \\ . & . & . & . & . & . & . & . & . \\ 0 & . & . & . & . & 0 & q & 0 & p \\ d_0 & d_1 & d_2 & . & . & . & . & . & d_b \end{pmatrix} \tag{3.5.12}$$

where $\{c_j\}$ and $\{d_j\}$ are probability distributions on $j = 0, 1, \ldots, b$. The nature of the barriers at states 0 and b are expressed by choosing $\{c_j\}$ and $\{d_j\}$ appropriately. Thus the barrier at 0 is *absorbing* if $c_0 = 1$, for then $p_{00} = 1$ and $p_{0k} = 0$ for $k = 1, 2, \ldots, b$. The barrier is *reflecting* if $c_0 < 1$, and there is then a positive probability of the particle being reflected back into the interior of the state space. This example is continued in Exercises 4–7. □

Exercises

1. Let $\mathscr{P} = (p_{ik}; i, k = 0, \pm 1, \pm 2, \ldots)$ be the transition matrix of a homogeneous Markov chain. From the identity

$$\mathscr{P}^{m+n} = \mathscr{P}^m \mathscr{P}^n$$

verify the *Chapman–Kolmogorov* equations

$$p_{ik}^{(n+m)} = \sum_j p_{ij}^{(n)} p_{jk}^{(m)}.$$

In the special case $p_{ii+1} = p, p_{ii-1} = q = 1-p, p_{ik} = 0$ otherwise, show that the transition probabilities satisfy the difference equations

$$\begin{aligned} p_{ik}^{(n+1)} &= qp_{i-1\,k}^{(n)} + pp_{i+1\,k}^{(n)} \\ p_{ik}^{(n+1)} &= pp_{ik-1}^{(n)} + qp_{ik+1}^{(n)}. \end{aligned} \tag{3.5.13}$$

2. Suppose the state space of a Markov chain is $\{0, 1\}$ and that the transition matrix is

$$\mathscr{P} = \begin{pmatrix} p_1 & q_1 \\ p_2 & q_2 \end{pmatrix}, \qquad p_1+q_1 = 1 = p_2+q_2.$$

Show that for $n = 0, 1, 2, \ldots$ (with $p_{ij}^{(0)} = \delta_{ij}$)

$$\begin{aligned} p_{00}^{(n+1)} &= p_1 p_{00}^{(n)} + p_2 p_{01}^{(n)} \\ p_{01}^{(n+1)} &= q_1 p_{00}^{(n)} + q_2 p_{01}^{(n)}. \end{aligned}$$

Use the generating functions

$$G_{00}(z) = \sum_{n=0}^{\infty} z^n p_{00}^{(n)} \quad \text{and} \quad G_{01}(z) = \sum_{n=0}^{\infty} z^n p_{01}^{(n)}$$

to find explicit expressions for $p_{00}^{(n)}$ and $p_{01}^{(n)}$.

3. A communications systems transmits the digits 0 and 1. Each digit transmitted must pass through three stages at each of which there is a probability p that an incoming digit will leave unchanged. What is the probability that a digit entering as 0 will be transmitted as 0?

4. (Continuation of Example 4.) In the transition matrix (3.5.12) take $c_0 = q$, $c_1 = p$ and $d_b = p$, $d_{b-1} = q$, implying that all other c_j, d_j vanish. As in Exercise 1 show that the transition probabilities for this restricted random walk satisfy.

$$\begin{aligned} p_{ik}^{(n+1)} &= qp_{(i-1)k}^{(n)} + pp_{(i+1)k}^{(n)}, \qquad i = 1, 2, \ldots, b-1, \\ p_{0k}^{(n+1)} &= qp_{0k}^{(n)} + pp_{1k}^{(n)}, \\ p_{bk}^{(n+1)} &= qp_{(b-1)k}^{(n)} + pp_{bk}^{(n)}. \end{aligned}$$

On the other hand if the barriers at 0 and b are absorbing ($c_0 = 1 = d_b$ in (3.5.12)) the transition probabilities satisfy

$$p_{ik}^{(n+1)} = qp_{(i-1)k}^{(n)} + pp_{(i+1)k}^{(n)}, \quad i = 2, 3, \dots, b-2,$$

$$p_{1k}^{(n+1)} = pp_{2k}^{(n)} + q\delta_{0k}, \quad p_{(b-1)k}^{(n+1)} = qp_{(b-2)k}^{(n)} + p\delta_{bk} \quad (3.5.14)$$

since $\quad p_{00}^{(n)} = 1, \quad p_{0k}^{(n)} = 0, \quad k = 1, 2, \dots, b,$

$\qquad p_{bb}^{(n)} = 1, \quad p_{bk}^{(n)} = 0, \quad k = 0, 1, \dots, b-1.$

Note that in each case the general equation is the same as the first equation of (3.5.13). The nature of the barriers at 0 and b impose different boundary conditions on this general equation.

5. Simple random walk constrained by absorbing barriers at 0 and b is a case of the so-called *gambler's ruin problem.* A gambler with initial fortune i plays a game which consists of a sequence of independent trials (e.g. tossing a coin) at each of which he wins a unit amount with probability p or loses a unit amount with probability $q = 1-p$. If his opponent's initial capital is $b-i$, the evolution of the gambler's fortune through time is described by a restricted random walk whose transition probabilities satisfy (3.5.14). The game stops when either the walk is absorbed at 0 and the gambler is ruined, or is absorbed at b and the opponent is ruined.

Let $v_i^{(n)}$ be the probability that the game stops at or before the nth trial, given intial capital i. Verify that

$$v_i^{(n)} = p_{i0}^{(n)} + p_{ib}^{(n)}$$

and hence, from (3.5.14), the $v_i^{(n)}$ satisfy

$$v_i^{(n+1)} = qv_{i-1}^{(n)} + pv_{i+1}^{(n)}, \qquad i = 2, 3, \dots, b-2$$

$$v_1^{(n+1)} = pv_2^{(n)} + q, \qquad v_{b-1}^{(n+1)} = qv_{b-2}^{(n)} + p.$$

The limit $\lim_{n\to\infty} v_i^{(n)} = v_i$ is the probability that the game stops in finite time. Show that $v_i = 1$. Hence for all finite b the game is certain to end in finite time.

6. (Continuation.) Suppose the gambler's opponent has infinite capital ($b = \infty$). The probability v_i that the game stops after a

finite number of trials is now the probability that the gambler will be ruined. Show that in this case we have the difference equations

$$v_i = qv_{i-1}+pv_{i+1}, \qquad i = 2,3,\ldots,$$
$$v_1 = q+pv_2,$$

with solution

$$v_i = \begin{cases} 1 & \text{if } p \leqslant q, \\ (q/p)^i & \text{if } p > q. \end{cases}$$

Conclude that ultimate ruin is certain if the game is fair ($p = q$) or unfavourable ($p < q$). However there is still the non zero probability $(q/p)^i$ of ruin even for a favourable game.

7. (Continuation.) Retaining the assumption $b = \infty$ let T_i denote the time to ruin for a gambler with initial capital i. T_i is the number of trials required for the Markov chain to reach state 0 from i for the first time and is a *first passage time.* Let $\Pr(T_i = n) = u_i^{(n)}$. Recognising that $u_i^{(n)} = v_i^{(n+1)} - v_i^{(n)}$ form the equations

$$u_i^{(n)} = qu_{i-1}^{(n-1)}+pu_{i+1}^{(n-1)}, \qquad i = 2,3,\ldots,$$
$$u_1^{(n)} = pu_2^{(n-1)}, \qquad u_1^{(1)} = q,$$
$$u_i^{(0)} = 0, \qquad i = 1,2,\ldots,$$
$$u_0^{(0)} = 1,$$

which are valid for $n = 0,1,2,\ldots$. Show that the generating functions $U_i(z) = \sum_{n=0}^{\infty} z^n u_i^{(n)}$ satisfy

$$U_i(z) = zqU_{i-1}(z)+zpU_{i+1}(z)$$
$$U_1(z) = zq+zpU_2(z)$$

with solution

$$U_i(z) = \left\{\frac{1-(1-4pqz^2)^{\frac{1}{2}}}{2pz}\right\}^i = \{U_1(z)\}^i.$$

By Abel's Theorem

$$\lim_{n\to\infty} \Pr(T_i \leqslant n) = \lim_{z\to 1-} U_i(z) = v_i = \begin{cases} 1 & \text{if p} \leqslant \text{q}, \\ (q/p)^i & \text{if } p > q. \end{cases}$$

Thus $\{\Pr(T_i = n), n = 1, 2, \ldots\}$ is a proper probability distribution only when $p \leqslant q$. In the case of a fair game $p = q = \frac{1}{2}$ and

$$U_1(z) = \{1-(1-z^2)^{\frac{1}{2}}\}/z$$

implying

$$U_i'(1) = ET_i = \infty.$$

An interesting conclusion is that if the game is fair ultimate ruin is certain but the expected time to ruin is infinite.

Chapter 4
WEAK CONVERGENCE

By the weak convergence of a sequence of functions is meant their convergence to a limit function at all points of continuity of the latter. The present chapter is concerned with sequences of distribution functions convergent in this sense. If the limit of such a sequence of distribution functions is itself a distribution function, then we have what is commonly called *convergence in law* or *convergence in distribution* of the corresponding sequence of random variables. (Definition 4.1.1).

The importance of this form of convergence lies in the fact that it enables one to approximate for large n the distribution function of the nth member of a sequence of random variables $\{X_n, n = 1, 2, \ldots\}$ convergent in distribution. In particular, estimates of probabilities such as $\Pr(a < X_n < b)$ can be obtained and these are of evident interest in applications. The fundamental result in this direction is the Central Limit Theorem 4.3.1 and this theorem explains to a large extent the frequent appearance of the normal distribution in the theory and applications of probability.

In this chapter *degenerate* and *dishonest* random variables appear for the first time. A degenerate random variable is one whose probability mass is concentrated at a single point.

$$\Pr(X = c) = 1, \quad \Pr(X = x) = 0, \qquad x \neq c.$$

Its distribution function is a step function with a single jump of height unity at the point $x = c$. On the other hand a dishonest (or improper or defective) random variable is such that its total probability mass is less than unity; that is

$$\lim_{x\to\infty} \Pr(X \leqslant x) < 1.$$

Dishonest random variables (whilst still functions from Ω into the real line) are not random variables in the sense in which the concept was introduced and developed in § 1.4. Nonetheless it is convenient to abuse terminology in this instance, since the degenerate and dishonest cases frequently occur as the limits in distribution of a sequence of ordinary, honest random variables. Examples are the weak law of large numbers (§ 4.2) in which the limit is a constant, and some first passage distributions for Markov chains (§ 4.5) which are dishonest.

The main sections of the chapter are the first three. The results in § 4.4 are important in statistical inference but are not of particular probabilistic significance. § 4.5 is concerned with certain aspects of the limit behaviour of the Markov chains introduced in § 3.5.

4.1 SEQUENCES OF DISTRIBUTION FUNCTIONS

Let $X_1, X_2, \ldots, X_n$ be a sequence of random variables with distribution functions $F_1(x), F_2(x), \ldots, F_n(x)$ respectively. The limit $\lim_{n\to\infty} F_n(x)$ may or may not exist, and even if it does may not be a distribution function.

Definition 4.1.1 If the sequence of distribution functions $\{F_n(x), n = 1, 2, \ldots\}$ converges to a limit function

$$\lim_{n\to\infty} F_n(x) = F(x) \tag{4.1.1}$$

at all points of continuity of $F(x)$, with $F(x)$ also a distribution function, then we say that the corresponding sequence of random variables $\{X_n\}$ *converges in distribution.* □

This form of convergence is also called *convergence in law*, meaning that the probability law (i.e. distribution function) of X_n converges to a function which is also a probability law. Recall that $F(x)$ is a distribution function if and only if it is non-decreasing and $F(-\infty) = 0$, $F(\infty) = 1$.

If indeed $\{X_n\}$ converges in distribution, the limit $F(x)$ of (4.1.1) induces a probability space on which can be defined a random variable X with

$$\Pr(X \leqslant x) = F(x).$$

This X, induced by $F(x)$, is called the *limit in distribution* of $\{X_n\}$, and we can say that the sequence $\{X_n\}$ converges in distribution to X. However, we cannot write

$$\lim_{n\to\infty} X_n = X \quad \text{or} \quad X_n \to X,$$

since we have not yet defined notions of convergence for the random variables themselves. Thus if $\{X_n\}$ does converge in distribution to X we will indicate when necessary this mode of convergence by writing

$$\text{d}\lim_{n\to\infty} X_n = X \quad \text{or} \quad X_n \xrightarrow{d} X.$$

The limit random variable is unique only in the sense that any other limit variable must have the same distribution function at all points of continuity.

Other modes of convergence will be discussed later in this chapter and in the next. One important point to note is that if $X_1, X_2, \ldots, X_n$ are defined on the product sample space

$$\begin{aligned} \Omega^{(n)} &= \Omega_1 \times \Omega_2 \times \ldots \times \Omega_n \\ &= \{(w_1, w_2, \ldots, w_n) \mid w_j \in \Omega_j, j = 1, 2, \ldots, n\}, \end{aligned}$$

discussion of the limit of the sequence $\{X_n\}$ strictly speaking involves considering the sample space $\Omega^{(\infty)}$ whose points are vectors of infinite dimension, since this is the space on which the infinite sequence of random variables is defined. Convergence in distribution side-steps these complications by defining the limit random variable through an induced space. It is perhaps the most useful (as well as the simplest) of the various modes of convergence as far as applications are concerned, since it enables us to obtain approximations for various probabilities of interest.

Examples are not hard to find. A simple case is

$$\begin{aligned} F_n(x) &= 1 - e^{-\lambda_n x}, \qquad 0 < x, \\ &\to 1 - e^{-\lambda x} \end{aligned}$$

provided $\lambda_n \to \lambda$ in the usual sense. Indeed one frequently has convergence not only of distribution functions but also of densities.

Example 1 Suppose each X_n is binomially distributed

$$\Pr(X_n = x) = \binom{n}{x}\left[\frac{\lambda}{n}\right]^x\left[1-\frac{\lambda}{n}\right]^{n-x}, \qquad x = 0, 1, 2, \ldots, n.$$

The special feature here is that the parameter $p = \lambda/n$ decreases to zero as n increases and because of this we can work directly with the sequence of probabilities $\{\Pr(X_n = x), n = 1, 2, \ldots\}$.

After some rearrangement,

$$\lim_{n\to\infty} \Pr(X_n = x) = \frac{\lambda^x}{x!}\lim_{n\to\infty}\frac{n(n-1)\ldots(n-x+1)}{n^x}\cdot\left[1-\frac{\lambda}{n}\right]^{n-x}$$

Since

$$\lim_{n\to\infty}\left[1-\frac{\lambda}{n}\right]^n = e^{-\lambda}$$

and

$$\lim_{n\to\infty} 1\left[1-\frac{1}{n}\right]\left[1-\frac{2}{n}\right]\ldots\left[1-\frac{(x+1)}{n}\right] = 1$$

for every finite x, we find

$$\lim_{n\to\infty} \Pr(X_n = x) = e^{-\lambda}\frac{\lambda^x}{x!}, \qquad x = 0, 1, 2, \ldots.$$

That is, this sequence of binomially distributed random variables converges in distribution to X, with X distributed according to the Poisson law. □

Example 2 Suppose $X_n \frown N(0, \sigma_n^2)$. Then

$$F_n(x) = \frac{1}{\sigma_n(2\pi)^{\frac{1}{2}}}\int_{-\infty}^{x} \exp\left[-\frac{1}{2}\left(\frac{y}{\sigma_n}\right)^2\right]dy$$

$$= \frac{1}{(2\pi)^{\frac{1}{2}}}\int_{-\infty}^{x/\sigma_n} \exp\left[-\frac{u^2}{2}\right]du$$

The convergence of $F_n(x)$ is completely determined by the behaviour of the sequence of standard deviations $\{\sigma_n\}$. If $\sigma_n \to \sigma$, a finite non-zero constant, then $X_n \xrightarrow{d} X$ with $X \frown N(0, \sigma^2)$. On the other hand if $\sigma_n \to 0$ then

$$\lim_{n\to\infty} F_n(x) = F(x) = \begin{cases} 0 & \text{if } x < 0 \\ \frac{1}{2} & \text{if } x = 0 \\ 1 & \text{if } x > 0 \end{cases}$$

Considering only points of continuity of the limit $F(x)$, namely $\{x \mid x < 0, x > 0\}$, we conclude that X_n converges in distribution to a *degenerate random variable* whose distribution has total probability mass 1 concentrated at a single point, in this case the origin. Furthermore for the sequence of density functions

$$\lim_{n\to\infty} \frac{e^{-\frac{1}{2}(y/\sigma_n)^2}}{\sigma_n(2\pi)^{\frac{1}{2}}} = \begin{cases} 0 & \text{if } y \neq 0, \\ \infty & \text{if } y = 0, \end{cases}$$

if $\sigma_n \to 0$. □

General results on what we have called convergence in distribution were first formulated during the eighteenth century for sequences of binomially distributed random variables. Recall from Example 3.3.4 that the sum

$$S_n = Y_1 + Y_2 + \ldots + Y_n$$

of n independent and identically distributed indicator variables is binomial with

$$\Pr(S_n = x) = \binom{n}{x} p^x (1-p)^{n-x}, \qquad x = 0, 1, 2, \ldots, n.$$

The parameter p is the constant probability of a 'head', $\Pr(Y_j = 1) = p$. It is instructive to consider the sequence $\{S_n\}$ in more detail.

Note firstly that $\lim_{n\to\infty} \Pr(S_n = x) = 0$ for every finite x, implying that the distribution function $\Pr(S_n \leqslant x)$ and indeed all probabilities of the form $\Pr(a < S_n \leqslant b)$, a, b, constant, also tend to zero as $n \to \infty$. The mean and variance of S_n are easily found to be

$$ES_n = np, \qquad \operatorname{Var} S_n = npq,$$

where $q = 1-p$, which are both unbounded in n. For large n we then have, as it were, a probability mass that is increasingly dispersed about a centre of gravity which itself is increasing unboundedly. This suggests that sensible limit laws will only result for quantities of the form $\Pr(a_n < S_n \leqslant b_n)$ with a_n, b_n chosen so as to keep pace with increasing mean and variance. The problem of finding suitable a_n and b_n can be resolved in several ways, the most common of which is to select these numbers to yield a transformation of S_n such that the transformed random variable has constant mean and variance.

A suitable transformation is the linear one

$$Z_n = \frac{S_n - ES_n}{(\mathrm{Var}\, S_n)^{\frac{1}{2}}} \tag{4.1.2}$$

which in the present case reduces to

$$Z_n = \frac{S_n - np}{(npq)^{\frac{1}{2}}}.$$

Z_n is called the *standardised* or *normalised* version of S_n and its distinctive feature is that uniformly in n,

$$EZ_n = 0, \qquad \mathrm{Var}\, Z_n = 1.$$

For the binomial variable S_n we have the identity for any finite $a < b$,

$$\Pr(a \leqslant Z_n \leqslant b) = \Pr(np + a(npq)^{\frac{1}{2}} \leqslant S_n \leqslant np + b(npq)^{\frac{1}{2}}).$$

We seek information about the limiting value of such probabilities for S_n through the corresponding probabilities for Z_n, whose mean and variance are independent of n.

In the binomial case the standard variable Z_n is, from (4.1.2), distributed over the $n+1$ points

$$\left\{\frac{j-np}{(npq)^{\frac{1}{2}}},\ j = 0, 1, 2, \ldots, n\right\}$$

with distribution

$$\Pr\left(Z_n = \frac{j-np}{(npq)^{\frac{1}{2}}}\right) = \Pr(S_n = j) = \binom{n}{j} p^j q^{n-j}, \quad j = 0, 1, \ldots, n.$$

Its moment generating function is (from the table on page 108)

$$\begin{aligned} E\exp(-sZ_n) &= E\exp\{-s(S_n - np)/(npq)^{\frac{1}{2}}\} \\ &= \exp\{s(np/q)^{\frac{1}{2}}\} E\exp\{-sS_n/(npq)^{\frac{1}{2}}\} \\ &= \exp\{s(np/q)^{\frac{1}{2}}\}[q + p\exp\{-s/(npq)^{\frac{1}{2}}]^n. \end{aligned} \tag{4.1.3}$$

Having this information about $\{Z_n\}$ our problem is to evaluate $\lim_{n\to\infty} \Pr(Z_n \leqslant x)$. Rather than attempting this* we argue heuristically as follows. Since there is a 1–1 correspondence between generating functions and distribution functions (Theorem 2.2.1(c) and 2.3.1) it seems a reasonable supposition that if a sequence of generating functions converges to a limit $M(s)$ which is itself a generating function, then the corresponding sequence of distribution functions converges to a distribution function whose generating function is $M(s)$.

It turns out that this is indeed true and in Exercise 5 you are asked to verify the convergence of Ee^{-sZ_n} in (4.1.3) to $e^{\frac{1}{2}s^2}$ and so recover the De Moivre–Laplace theorem. Since characteristic functions exist for all random variables we state the result in full generality as follows:

Theorem 4.1.1 (Continuity Theorem). Let $F_n(x)$ and $\psi_n(s) = Ee^{isX_n}$ be respectively the distribution function and characteristic function of X_n. The sequence $\{F_n(x)\}$ converges weakly to a limiting distribution function $F(x)$ if and only if $\psi_n(s)$ converges for every s to a limit $\psi(s)$ which is continuous at $s = 0$. When this holds $\psi(s)$ is the characteristic function of $F(x)$. □

The name is appropriate because the theorem gives conditions under which the 1–1 correspondence between distribution functions and characteristic functions is continuous in the sense of being preserved under limits.

The importance of the theorem lies in the fact that it is frequently easier to find the limits of sequences of characteristic functions or other generating functions than of the original distribution functions, and it therefore provides a useful method for verifying convergence in distribution. This is particularly true when considering sums of independent random variables for then convolutions are transformed into products. (4.1.3) is a case in point, and the proof of the central limit theorem in § 4.3 is another. For other sequences a direct approach may be preferable as the next example shows.

* That $\lim_{n\to\infty} \Pr(a < Z_n \leqslant b) = (2\pi)^{\frac{1}{2}} \int_a^b e^{-\frac{1}{2}x} \, dx$ is the assertion of the classical *De Moivre–Laplace* limit theorem. For a direct proof which uses Stirling's formula (2.5.15) to approximate binomial coefficients see FELLER I, Chapter 7.

Example 3 $X_1, X_2, \ldots, X_n$ are independent with common distribution function $F(x)$. The largest one is

$$M_n = \max(X_1, X_2, \ldots, X_n).$$

Since $M_n \leqslant x$ if and only if all of $X_1, X_2, \ldots, X_n$ are $\leqslant x$,

$$\Pr(M_n \leqslant x) = (F(x))^n = \{1-(1-F(x))\}^n.$$

As $n \to \infty$ the limit of $\Pr(M_n \leqslant x)$ is zero for all x for which $F(x) < 1$ and non-trivial limit laws only result when x increases with n in a suitable way. The characteristic function of M_n,

$$Ee^{isM_n} = \int_{-\infty}^{\infty} e^{isx} n(F(x))^{n-1} f(x)\, dx,$$

is usually complicated and explicit information on the rate of approach to zero of $\lim_{x\to\infty} (1-F(x))$ is more useful. A particular case appears in Exercise 13. □

A proof of the continuity theorem is not given here since the proofs available use in an essential way either Fourier inversion formulae and/or Stieltjes integrals, and this knowledge is not assumed. A major reason for the complicated nature of the argument is that the limits of distribution functions of discrete random variables (which are step functions) may be continuous and a general theory of integration covering both cases is necessary. An example is provided by the standardised binomial variable Z_n which converges in distribution to a (continuous) $N(0, 1)$ variable. Example 2 illustrates that discrete limits may also occur. We therefore take the continuity theorem as given. It is used extensively in this chapter with the result that sequences of distribution functions will often be studied through the corresponding sequence of characteristic functions.

Exercises

1. X_n is negatively binomial distributed with

$$\Pr(X_n = k) = \binom{n+k-1}{k} p^n (1-p)^k, \qquad k = 0, 1, 2, \ldots$$

Show that if $p \to 1$ as $n \to \infty$ in such a way that

$$\lim_{n\to\infty} n(1-p) = \mu$$

then

$$\lim_{n\to\infty} \Pr(X_n = k) = \frac{e^{-\mu}\mu^k}{k!}, \qquad k = 0, 1, 2, \ldots .$$

[*Hint*: $\lim_{n\to\infty} \{1+an^{-1}+o(n^{-1})\}^n = e^a$]

2. Use generating functions and the continuity theorem to verify the convergence in distribution to the Poisson of the random variables of Example 1 and Exercise 1.

3. The probability that a patient succumbs to a particular disease is 0·1. Of a group of 50 patients what is the probability that at least 49 survive? Compare numerically the answers given by the bionomial distribution and the Poisson approximation.

4. Find an approximation to the probability that of 500 people exactly one has a birthday on Christmas day. (See FELLER, I, page 155, for numerical comparisons).

5. Evaluate the limit as $n \to \infty$ in (4.1.3) to establish the De Moivre–Laplace theorem

$$\lim_{n\to\infty} \Pr\left(a \leqslant \frac{S_n - np}{(npq)^{\frac{1}{2}}} \leqslant b\right) = \frac{1}{(2\pi)^{\frac{1}{2}}}\int_a^b e^{-\frac{1}{2}x^2}\,dx.$$

6. Find the probability (approximately) that in 1000 tosses of a fair coin the number of heads lies between 480 and 520.

7. A fair die is rolled 60 times. Compare the probabilities given by the Poisson and Normal approximations for the event that the number of sixes lies between 10 and 30.

8. How many tosses of a fair coin are required to ensure that the probability is about 0·95 that at least 20 heads are obtained? (Use the fact that if $Z \frown N(0, 1)$ then $0{\cdot}05 = \Pr(Z \geqslant 1{\cdot}645)$).

9. Use characteristic functions to show that

$$\lim_{n\to\infty} \int_0^x \frac{e^{-n\lambda y} y^{n-1} (n\lambda)^n}{n!} dy = 1, \qquad \text{all } x > 0.$$

10. The Continuity Theorem 4.1.1 requires the limit $\psi(s)$ of the sequence of characteristic functions to be continuous at the origin. Show that this condition cannot be relaxed by examining the sequence of distribution functions

$$F_n(x) = \begin{cases} 0 & \text{if } x < -n, \\ \tfrac{1}{2}(1+xn^{-1}) & \text{if } -n \leqslant x \leqslant n, \\ 1 & \text{if } n < x, \end{cases}$$

and the corresponding sequence of characteristic functions.

11. Does convergence in distribution imply convergence of moments to the moments of the limit random variable? The answer is in general no, since the convergence of a sequence of characteristic functions $\psi_n(s)$ to $\psi(s)$ does not necessarily imply the convergence of the derivatives $\psi_n'(0)$ to $\psi'(0)$. However, prove the following for the stated class of random variables: If

$$G_n(z) = \sum_x z^x \Pr(X_n = x), \qquad n = 1, 2, \ldots,$$

is a sequence of generating functions with radius of convergence $R \geqslant 1+\delta$, where $\delta > 0$, then the convergence of $G_n(z)$ to a proper probability generating function implies the convergence of the moments EX_n^k to the moments of the limit random variable.

12. (Continuation.) On the other hand it is possible for $\{X_n\}$ to converge in distribution to a limit variable which has moments of all orders whereas EX_n^k, $n = 1, 2, \ldots$, need not exist for any $k \geqslant 1$. As an example consider the distribution functions

$$\Pr(X_n \leqslant x) = x, \qquad 0 \leqslant x \leqslant 1-n^{-1},$$

$$= 1 - \frac{1}{n\{2+n(x-1)\}}, \qquad 1-n^{-1} < x < \infty$$

13. (Continuation of Example 3.) If X is a positive variable and

$$\lim_{x\to\infty} x^3 \Pr(X > x) = 1$$

show that

$$\lim_{n\to\infty} \Pr(n^{-1/3} M_n \leqslant x) = \exp\{-x^{-3}\}, \quad x > 0,$$

$$= 0, \quad \text{otherwise.}$$

What is the limit distribution if $F(x) = 1 - e^{-x}$, $x > 0$?

4.2 THE WEAK LAW OF LARGE NUMBERS

The popular version of the 'law of averages' is that the arithmetic mean of repeated measurements of a quantity subject to error approximates more closely the true value as the number of observations increase. In probabilistic terms this translates as the assertion that if

$$S_n = Y_1 + Y_2 + \ldots + Y_n$$

is the sum of n independent and identically distributed random variables with expectation μ, then for a large n the sample average $n^{-1}S_n$ takes values 'close' to μ in the sense that the probability $\Pr(|n^{-1}S_n - \mu| \geqslant \varepsilon)$, is negligible, where ε is an arbitrary positive number.

The formal statement is provided by the classical form of the weak law of large numbers:

Theorem 4.2.1 If S_n is the sum of n independent and identically distributed random variables with mean μ, then for every $\varepsilon > 0$

$$\lim_{n\to\infty} \Pr(|n^{-1}S_n - \mu| \geqslant \varepsilon) = 0. \qquad \square$$

Note that $E(n^{-1}S_n) = n^{-1}n\mu = \mu$ and if $\sigma^2 < \infty$ is the common variance of the summands, then also

$$\mathrm{Var}\left[\frac{S_n}{n}\right] = \frac{n\sigma^2}{n^2} = \frac{\sigma^2}{n}.$$

In the case of finite variance we apply Chebyshev's inequality (1.5.4) to find

$$\Pr\left(\left|\frac{1}{n}S_n - \mu\right| \geqslant \frac{\delta\sigma}{\sqrt{n}}\right) \leqslant \frac{1}{\delta^2}$$

for every $\delta > 0$. Taking $\delta = \sigma^{-1}/n$ we have

$$\Pr(|n^{-1}S_n - \mu| \geqslant \varepsilon) \leqslant \sigma^2/\varepsilon^2 n \qquad (4.2.1)$$
$$\to 0$$

as $n \to \infty$ for every $\varepsilon > 0$.

Thus, given $\sigma^2 < \infty$, the weak law of large numbers is an immediate consequence of Chebyshev's inequality. The result was first proved* for sums of independent and identically distributed indicator variables whose sum (discussed in § 4.1) is binomially distributed,

$$\Pr(S_n = x) = \binom{n}{x} p^x(1 - p)^{n-x}. \qquad (4.2.2)$$

In this case $\mu = p$, $\sigma^2 = pq$ and from Theorem 4.2.1 we infer

$$\lim_{n\to\infty} \Pr(n(p-\varepsilon) \leqslant S_n \leqslant n(p+\varepsilon)) = 1.$$

This is, of course, a considerably weaker statement than

$$\lim_{n\to\infty} \Pr(np + a(npq)^{\frac{1}{2}} \leqslant S_n \leqslant np + b(npq)^{\frac{1}{2}}) = \Phi(b) - \Phi(a)$$

which you were asked to establish using the Continuity Theorem in Exercise 4.1.5. However, Theorem 4.2.1 can be shown to hold under the assumption of finite first moment only and is therefore valid for a much wider class of variables than those satisfying convergence to the normal law.

The simplest general proof of the Theorem uses the Continuity Theorem 4.1.1. Suppose firstly that

$$\psi(s) = Ee^{isY}$$

is the characteristic function of Y and suppose $\mu = EY$ is finite.

* Apparently by JACOB BERNOULLI (1654–1705) whose book *Ars Conjectaandi* appeared in 1713. Independent trials, outcomes of which can be one of two alternatives, are frequently called *Bernoulli trials.* We have described such trials in terms of sequences of indicator variables but will occasionally use the other notation.

From (2.3.7) and Theorem 2.3.9

$$\mu = \frac{1}{i}\frac{d}{ds}(\psi(s))\bigg|_{s=0}$$

and, as $s \to 0$,

$$\psi(s) = 1 + is\mu + o(s). \tag{4.2.3}$$

If $S_n = Y_1 + Y_2 + \ldots + Y_n$, with the Y_j independently and identically distributed as Y, it follows from Example 2.3.2 that

$$\begin{aligned} e^{is(S_n n^{-1} - \mu)} &= e^{-is\mu} E e^{isS_n/n} \\ &= e^{-is\mu}\left[\psi\left(\frac{s}{n}\right)\right]^n \\ &= e^{-is\mu}\left[1 + \frac{is}{n}\mu + o\left(\frac{s}{n}\right)\right]^n \\ &\to e^{-is\mu}e^{is\mu} = 1 \quad \text{as } n \to \infty. \end{aligned}$$

Appealing to Theorem 4.1.1 we have that the sequence

$$n^{-1}S_n - \mu, \qquad n = 1, 2, \ldots$$

converges in distribution to the random variable degenerate at the origin; that is,

$$\lim_{n\to\infty} \Pr(|n^{-1}S_n - \mu| < \varepsilon) = 1$$

for every $\varepsilon > 0$. This is equivalent to the assertation of Theorem 4.2.1, the proof of which is now complete.

A perusal of the preceding argument indicates that a crucial role is played by the Taylor expansion (4.2.3). Indeed, replacing μ by $i^{-1}\psi'(0)$ we have in fact proved the more general proposition:

Theorem 4.2.2 If S_n is the sum of n independent and identically distributed variables whose common characteristic function $\psi(s)$ is differentiable at $s = 0$, then for every $\varepsilon > 0$

$$\lim_{n\to\infty} \Pr\left(\left|\frac{1}{n}S_n - \frac{1}{i}\psi'(0)\right| \geqslant \varepsilon\right) = 0. \qquad \square$$

The additional generality lies in the fact that whilst the existence of a first moment implies that $\psi(s)$ is differentiable at the origin, (Theorem 2.3.8), the converse is not true (a counter example appears in Exercise 5.4.7). Theorem 4.2.2 therefore asserts a weak law of large numbers for variables without expectation. Necessary and sufficient conditions* are stated in Theorem 5.4.6 and the relevance here of these remarks is that limit theorems for averages may hold under rather general conditions.

Definition 4.2.1 Suppose $Y_1, Y_2, \ldots,$ are any random variables with partial sums $S_n = Y_1 + Y_2 + \ldots + Y_n$, $n = 1, 2, \ldots$. The sequence $\{Y_n\}$ is said to obey the *weak law of large numbers* if constants $b_n, n = 1, 2, \ldots,$ can be found such that for every $\varepsilon > 0$

$$\lim_{n \to \infty} \Pr(|n^{-1}S_n - b_n| \geqslant \varepsilon) = 0. \qquad \square$$

If first moments exist then it is natural to take $b_n = n^{-1}ES_n$. That independence is not required is shown by the following example which is continued in Exercise 7.

Example 1 Let $S_n = Y_1 + Y_2 + \ldots + Y_n$ where $EY_j = \mu_j$, $\text{Var } Y_j = \sigma_j^2$, and the Y_j are *not* independent. Then

$$ES_n = \sum_{j=1}^{n} \mu_j$$

$$\text{Var } S_n = \sum_{j=1}^{n} \sigma_j^2 + 2 \sum_{k=2}^{n} \sum_{j=1}^{k} \text{Cov}(X_j, X_k),$$

and by Chebyshev's inequality

$$\Pr(|S_n - \sum_{j=1}^{n} \mu_j| \geqslant \delta\sqrt{\text{Var } S_n}) \leqslant 1/\delta^2,$$

which is (putting $\varepsilon = \delta n^{-1}\sqrt{\text{Var } S_n}$)

$$\Pr\left(\left|\frac{1}{n}S_n - \frac{1}{n}\sum_{j=1}^{n} \mu_j\right| \geqslant \varepsilon\right) \leqslant \frac{\text{Var } S_n}{\varepsilon^2 n^2}.$$

* These involve technicalities the details of which are not important for present purposes.

Provided the variances and covariances of the Y_j are such that $\lim_{n\to\infty} n^{-2}\text{Var}\, S_n = 0$ we have

$$\lim_{n\to\infty} \Pr\left(\left|\frac{1}{n}S_n - \frac{1}{n}\sum_{j=1}^{n} \mu_j\right| \geqslant \varepsilon\right) = 0$$

for every $\varepsilon > 0$. □

It is apparent that the mode of convergence prescribed by the weak law may apply to random variables other than averages. This circumstance leads to the introduction of a new term.

Definition 4.2.2 Suppose $\{X_n\}$ is a sequence of random variables and that there is a random variable X such that the joint distribution of X_n and X is defined for every $n = 1, 2, \ldots$. If

$$\lim_{n\to\infty} \Pr(|X_n - X| \geqslant \varepsilon) = 0$$

for every $\varepsilon > 0$, then $\{X_n\}$ is said to *converge in probability* to X, written

$$X_n \xrightarrow{\text{p}} X.$$ □

Thus another way of stating Theorem 4.2.1 is that the sequence $\{n^{-1}S_n\}$ converges in probability to μ.

Whilst convergence in probability and convergence in distribution are equivalent if the limit variable is degenerate, it is not generally true that $X_n \xrightarrow{\text{d}} X$ implies $X_n \xrightarrow{\text{p}} X$. An obvious counter example is when $X_1, X_2, \ldots,$ are independently and identically distributed as X. For then one trivially has convergence in distribution to X since all the distribution functions $\Pr(X_n \leqslant x)$ are identical. On the other hand each $X_n - X_{n+m}$ is distributed as the symmetrised version of X and so is non-degenerate. The general result is that convergence in probability implies convergence in distribution but not conversely.

Theorem 4.2.3 If $X_n \xrightarrow{\text{p}} X$ then $X_n \xrightarrow{\text{d}} X$.

Proof. For every $\varepsilon > 0$

$$\begin{aligned}\Pr(X_n \leqslant x) &= \Pr(X_n \leqslant x, X \leqslant x+\varepsilon) + \Pr(X_n \leqslant x, X > x+\varepsilon)\\ &\leqslant \Pr(X \leqslant x+\varepsilon) + \Pr(X_n \leqslant x, X > x+\varepsilon)\\ &\leqslant \Pr(X \leqslant x+\varepsilon) + \Pr(|X_n - X| > \varepsilon).\end{aligned}$$

Interchanging the roles of X_n and X and replacing x by $x-\varepsilon$ we find by the same argument

$$\Pr(X \leqslant x-\varepsilon)-\Pr(|X_n-X| > \varepsilon) \leqslant \Pr(X_n \leqslant x).$$

If $n \to \infty$ in both inequalities it follows from Definition 4.2.2 that

$$\Pr(X \leqslant x-\varepsilon) \leqslant \lim_{n\to\infty} \Pr(X_n \leqslant x) \leqslant \Pr(X \leqslant x+\varepsilon).$$

But $\varepsilon > 0$ can be chosen arbitrarily small and hence

$$\lim_{n\to\infty} \Pr(X_n \leqslant x) = \Pr(X \leqslant x)$$

at every point of continuity. □

It has been noted already that convergence in distribution and convergence in probability are identical if the limit is a degenerate variable. For certain sequences we find the same equivalence even if the limit is not necessarily a constant, and in these special circumstances Theorem 4.2.3 does have a converse. For example

Theorem 4.2.4 For each $n = 1,2,\ldots,$ let $S_n = Y_1 + Y_2 + \ldots + Y_n$ where the Y_j are independent. If $S_n \xrightarrow{d} X$ then also $S_n \xrightarrow{p} X$.

Proof. By independence

$$Ee^{isS_n} = \prod_{j=1}^{n} Ee^{isY_j}$$

which by the continuity theorem converges to Ee^{isX} as $n \to \infty$. Then

$$\begin{aligned} E\exp\{is\sum_{j=n+1}^{n+m} Y_j\} &= \prod_{j=n+1}^{n+m} Ee^{isY_j} \\ &= \left(\prod_{j=1}^{n+m} Ee^{isY_j}\right)\Big/\left(\prod_{j=1}^{n} Ee^{isY_j}\right) \\ &\to 1 \text{ as both } n,m \to \infty. \end{aligned}$$

Therefore

$$\sum_{j=n+1}^{n+m} Y_j = S_{n+m} - S_n$$

converges to zero in both distribution and probability. In particular

$$\sum_{j=n+1}^{\infty} Y_j \xrightarrow{P} 0$$

as $n \to \infty$ and hence $S_n = \sum_{j=1}^{n} Y_j$ converges in probability. By the preceding theorem its limit in distribution must be X.

Example 2 Let $Y_1, Y_2, \ldots,$ be independent with $Y_j \frown N(0, \sigma_j^2)$. Then $Ee^{isY_j} = e^{-\frac{1}{2}s^2\sigma_j^2}$ and for the partial sum

$$E \exp(isS_n) = \prod_{j=1}^{n} \exp(-\tfrac{1}{2}s^2\sigma_j^2) = \exp(-\tfrac{1}{2}s^2 \sum_{j=1}^{n} \sigma_j^2)$$

Provided $\sum_{j=1}^{\infty} \sigma_j^2 = \sigma^2 < \infty$, S_n converges in distribution to a $N(0, \sigma^2)$ variable, and by the theorem, this convergence holds also in probability. □

Example 3 Behaviour of a very erratic nature is possible for the averages of certain variables without expectation. A case in point arises when Y is Cauchy with density

$$f(y) = \frac{1}{\pi(1+y^2)}, \qquad -\infty < y < \infty.$$

From the table on page 110

$$Ee^{isY} = e^{-|s|},$$

and hence also

$$E \exp(isn^{-1}S_n) = \{E \exp(isn^{-1}Y)\}^n = \exp(-|s|)$$

That is, the average has precisely the same distribution as any individual summand no matter how large n is taken, and the weak law does not apply. The Cauchy distribution provides an example where repeated averaging does not lead to increasing concentration about some central value. Trivially $\{n^{-1}S_n\}$ converges to Y in distribution but not in probability. Note also that the characteristic function $e^{-|s|}$ is not differentiable at the origin and (c.f. Theorem 5.4.6) $\lim_{y \to \infty} y \Pr(|Y| \geq y)$ is non-zero. □

Exercises

1. Verify by direct calculation that the weak law of large numbers holds for a sequence of independent $N(\mu, \sigma^2)$ variables.

2. To what constant does the average sum of squares of n independent variables converge in probability if their common distribution is

   (a) Poisson
   (b) negative binomial
   (c) gamma.

3. In a sequence of independent games it is possible to win or lose a unit stake at the jth game with probabilities p_j and $1-p_j$ respectively. Let S_n denote the total gain in the first n games. Show that

$$\lim_{n\to\infty} \Pr(n^{-1}|S_n - ES_n| \geqslant \varepsilon) = 0$$

for every $\varepsilon > 0$ irrespective of the values of the p_j.

4. (Continuation.) If $p_j = \frac{1}{2}+\delta_j$ with $\Sigma\,\delta_j < \infty$ then

$$n^{-1}S_n \xrightarrow{\text{p}} 0.$$

5. (Continuation.) Strengthen the weak law to show that for any p_j

$$\lim_{n\to\infty} \Pr(|S_n - ES_n| \geqslant \varepsilon n^{\frac{1}{2}} \log n) = 0$$

6. (Continuation.) If $p_j = p_1, j$ odd, and $p_j = p_2, j$ even, then

$$n^{-1}S_n \xrightarrow{\text{p}} 2^{-1}(p_1 + p_2 - 1).$$

Does $(S_n - ES_n)/\sqrt{\text{Var } S_n}$ converge to the normal law?

7. Prove that the weak law holds for the sequence in Example 1 provided the variances are bounded and $\lim\limits_{|j-k|\to\infty} \text{Cov}(X_j, X_k) = 0$.

8. Modify the proof of Theorem 4.2.3 to establish that if $X_n - Y_n \xrightarrow{\text{p}} 0$ and $X_n \xrightarrow{\text{d}} X$ then $Y_n \xrightarrow{\text{d}} X$.

9. $Y_1, Y_2, \ldots,$ are independent with

$$\Pr(Y_n = b_n) = p_n, \quad \Pr(Y_n = -b_n) = p_n, \quad \Pr(Y_n = 0) = 1-2p_n.$$

Show that $\{Y_n\}$ obeys the weak law of large numbers if

$$n^{-2} \sum_{j=1}^{n} p_j b_j^2 \to 0$$

as $n \to \infty$.

10. Show that the weak law holds for all sequences of uniformly bounded random variables.

11. *Mean square convergence.* $\{X_n\}$ is said to converge to X in mean square if $\lim_{n\to\infty} E(X_n - X)^2 = 0$. Use Chebyshev's inequality to prove that mean square convergence implies convergence in probability. The converse is not true. For a counter example consider nonnegative variables with

$$\Pr(X_n \leqslant x) = 1-(1+nx)^{-1}, \qquad x \geqslant 0.$$

$X_n \xrightarrow{\text{P}} 0$ but all expectations are infinite. This example shows incidentally that convergence in probability implies nothing about the convergence of moments.

4.3 THE CENTRAL LIMIT THEOREM

The De Moivre–Laplace limit theorem of § 4.1 asserts the convergence in distribution of the standardised binomial variables $\{(S_n - np)(npq)^{-\frac{1}{2}}\}$ to a $N(0,1)$ variable. In Exercise 4.1.5 it was required to prove this via the continuity theorem by showing that the sequence of moment generating functions converged to $\exp(\frac{1}{2}s^2)$, the moment generating function of the standard normal distribution.

The argument used in the binomial case generalises without difficulty to any sequence of partial sums of independent and identically distributed random variables with moment generating functions. Suppose

$$S_n = Y_1 + Y_2 + \ldots + Y_n$$

is the general term of such a sequence and let the mean and variance of Y be respectively μ and σ^2. If $M(s) = Ee^{-sY}$ is the common moment generating function of the Y_j then for all s in the interval of convergence

$$M(s) = 1 + \sum_{k=1}^{\infty} \frac{(-s)^k}{k!}\mu_k$$

where $\mu_k = EY^k$. In particular (since $\mu_2 = \sigma^2 + \mu^2$)

$$M(s) = 1 - s\mu + s^2\mu_2/2! + o(s^2) \qquad \text{as } s \to 0. \tag{4.3.1}$$

If $Z_n = (S_n - n\mu)(\sigma\sqrt{n})^{-1}$ it follows as in the binomial case that

$$\begin{aligned} E\exp(-sZ_n) &= \exp\left[\frac{s\mu\sqrt{n}}{\sigma}\right] E\exp\left[-\frac{s}{\sigma\sqrt{n}}S_n\right] \\ &= \exp\left[\frac{s\mu\sqrt{n}}{\sigma}\right]\left[M\left(\frac{s}{\sigma\sqrt{n}}\right)\right]^n \\ &= \exp\left[\frac{s\mu\sqrt{n}}{\sigma}\right]\left[1 - \frac{s\mu}{\sigma\sqrt{n}} + \frac{s^2\mu_2}{2!\sigma^2 n} + o\left(\frac{s^2}{\sigma^2 n}\right)\right]^n. \end{aligned}$$

For n sufficiently large,

$$\begin{aligned} \log Ee^{-sZ_n} &= s\frac{\mu}{\sigma}\sqrt{n} + n\log\left[1 - \frac{s\mu}{\sigma\sqrt{n}} + \frac{s^2\mu_2}{2\sigma^2 n} + o\left(\frac{s^2}{\sigma^2 n}\right)\right] \\ &= s\frac{\mu}{\sigma}\sqrt{n} - n\left[\frac{s\mu}{\sigma\sqrt{n}} - \frac{s^2\mu_2}{2\sigma^2 n} + o\left(\frac{s^2}{\sigma^2 n}\right)\right. \\ &\qquad \left. + \frac{1}{2}\left\{\frac{s\mu}{\sigma\sqrt{n}} - \frac{s^2\mu_2}{2\sigma^2 n} + o\left(\frac{s^2}{\sigma^2 n}\right)\right\}^2 + \dots\right], \end{aligned}$$

since $\log(1-z) = -\{z + \frac{1}{2}z^2 + \frac{1}{3}z^3 + \dots\}$ when $|z| < 1$. Regrouping terms,

$$\begin{aligned} \log Ee^{-sZ_n} &= n\left[\frac{s^2\mu_2}{2\sigma^2 n} - \frac{s^2\mu^2}{2\sigma^2 n} + o\left(\frac{s^2}{\sigma^2 n}\right)\right] \\ &= \frac{s^2}{2} + n.o\left(\frac{s^2}{\sigma^2 n}\right), \qquad \text{since } \mu_2 = \sigma^2 + \mu^2, \\ &\to \tfrac{1}{2}s^2 \quad \text{as } n \to \infty. \end{aligned}$$

Hence

$$\lim_{n\to\infty} Ee^{-sZ_n} = e^{\frac{1}{2}s^2},$$

the moment generating function of the $N(0,1)$ distribution, and from Theorem 4.1.1,

$$\lim_{n\to\infty} \Pr(Z_n \leqslant x) = \frac{1}{(2\pi)^{\frac{1}{2}}} \int_{-\infty}^{x} e^{-\frac{1}{2}y^2}\, dy. \tag{4.3.2}$$

This proof of (4.3.2) assumed the existence of a moment generating function Ee^{-sY}, implying in particular that all moments are finite (Theorem 2.3.3). Scrutiny of the argument makes clear that only the Taylor expansion (4.3.1) is required and for this a finite variance is sufficient (Theorem 2.3.9). With Ee^{isZ_n} in place of Ee^{-sZ_n} only trivial alterations are required and we have therefore established the fundamental result known as the Central Limit Theorem:

Theorem 4.3.1 If S_n is the sum of n independent and identically distributed variables with mean μ and variance σ^2, then for any $a < b$,

$$\lim_{n\to\infty} \Pr\left(a \leqslant \frac{S_n - n\mu}{\sigma\sqrt{n}} \leqslant b\right) = \Phi(b) - \Phi(a)$$

where

$$\Phi(x) = \frac{1}{(2\pi)^{\frac{1}{2}}} \int_{-\infty}^{x} e^{-\frac{1}{2}y^2}\, dy. \qquad \square$$

The weak law Theorem 4.2.1 supplied the estimate

$$\lim_{n\to\infty} \Pr(n\mu - n\varepsilon \leqslant S_n \leqslant n\mu + n\varepsilon) = 1$$

for all $\varepsilon > 0$ provided only the first moment μ was guaranteed. Under the additional assumption of finite variance Theorem 4.3.1 is a considerable sharpening of the weak law since, taking $a = -\varepsilon/\sigma$, $b = \varepsilon/\sigma$, it states that

$$\lim_{n\to\infty} \Pr(n\mu - \varepsilon\sqrt{n} \leqslant S_n \leqslant n\mu + \varepsilon\sqrt{n}) = \Phi\left(\frac{\varepsilon}{\sigma}\right) - \Phi\left(\frac{-\varepsilon}{\sigma}\right).$$

We can therefore estimate for large n the probability that S_n takes values in an interval of length $2\varepsilon\sqrt{n}$ about its mean. More generally an important application of the central limit theorem is to approximate quantities such as $\Pr(a \leqslant S_n \leqslant b)$ for n finite but large as in Exercises 4.1.6–8 and 4.3.1, 2.

If more information about the distribution function $\Pr(Y \leqslant y)$ is available it may be possible to evaluate the error of these approximations. Various estimates have been found but we state without proof only the most general result, which comes in the form of a uniform upper bound.

Theorem 4.3.2 (Berry–Esséen) If $E|Y-\mu|^3 < \infty$ then for all x

$$\left|\Pr\left(\frac{S_n - n\mu}{\sigma\sqrt{n}} \leqslant x\right) - \Phi(x)\right| \leqslant \frac{c}{\sigma^3\sqrt{n}} E|Y-\mu|^3.$$

c is a universal constant calculated to lie between 2 and 3. □

The discussion has been confined to sums of independent and identically distributed variables. For more general sequences the method used to obtain Theorem 4.3.1 fails since the normalised sum $(S_n - ES_n)/\sqrt{\operatorname{Var} S_n}$ no longer necessarily has a characteristic function that is the nth power of another characteristic function. However it turns out that convergence in distribution to the normal continues to hold for a remarkably wide class of random variables and for this reason the normal law plays a central role in probability and statistics.

Definition 4.3.1 Let $S_n = X_1 + X_2 + \ldots + X_n$. The sequence $\{X_n\}$ is said to satisfy the central limit theorem if for every x

$$\lim_{n\to\infty} \Pr\left(\frac{S_n - ES_n}{\sqrt{\operatorname{Var} S_n}} \leqslant x\right) = \frac{1}{(2\pi)^{\frac{1}{2}}} \int_{-\infty}^{x} e^{-\frac{1}{2}y^2}\, dy.$$ □

We have established the central limit theorem only in the relatively simple case of independent identically distributed variables with finite variances. A solution to the problem for independent but not necessarily identically distributed variables is provided by the Lindberg–Feller theorem stated in Exercise 10.

Turning* to another aspect of Theorem 4.3.1, note that it concerns the convergence of distribution, not density, functions to the normal. Partly this is due to stating the theorem in terms wide enough to cover both continuous and discrete variables, but also certain other analytical difficulties have to be overcome to establish 'local' limit theorems, i.e. limit theorems for densities or individual probabilities. Indeed the continuity theorem as stated in Theorem 4.1.1 refers specifically to distribution functions and not to sequences of densities.

When the summands have a density it is true that (except for an unusually badly behaved class of random variables)

$$\lim_{n\to\infty} \frac{d}{dx} \Pr\left[\frac{S_n - n\mu}{\sigma\sqrt{n}} \leqslant x\right] = \frac{1}{(2\pi)^{\frac{1}{2}}} e^{-\frac{1}{2}x^2}. \tag{4.3.3}$$

Discussion of (4.3.3) is omitted here but we consider the discrete case in some detail.

If Y is distributed over the points $a+kh$, $k = 0, \pm 1, \pm 2, \ldots$, with a and $h > 0$ fixed, the normalised variable $Z_n = (S_n - n\mu)/(\sigma\sqrt{n})$ has non-zero probabilities for values only in the set

$$\frac{a-n\mu}{\sigma\sqrt{n}} + k\frac{h}{\sigma\sqrt{n}}, \qquad k = 0, \pm 1, \pm 2, \ldots \tag{4.3.4}$$

It is therefore futile to ask for the limit of $\Pr(Z_n = x)$ unless x is of this form. For these x the inversion formula for lattice variables, Theorem 2.3.10, yields

$$\Pr(Z_n = x) = \frac{h}{2\pi\sigma\sqrt{n}} \int_{-\pi\sigma\sqrt{n}/h}^{\pi\sigma\sqrt{n}/h} e^{-isx}(Ee^{isZ_n})\, ds.$$

Then, from Exercise 2.3.15

$$\frac{\sigma\sqrt{n}}{h} \Pr(Z_n = x) - \frac{1}{\sqrt{2\pi}} e^{-\frac{1}{2}x^2} = \frac{1}{2\pi} \int_{-\pi\sigma\sqrt{n}/h}^{\pi\sigma\sqrt{n}/h} e^{-isx}(Ee^{isZ_n} - e^{-\frac{1}{2}s^2})\, ds$$

$$- \frac{1}{2\pi} \int_{|s| > \pi\sigma\sqrt{n}/h} e^{-isx - \frac{1}{2}s^2} ds. \tag{4.3.5}$$

* The remainder of this section treats a more specialised and difficult topic and could be deferred until a later reading.

If the absolute value of the right hand side of (4.3.5) can be shown to approach zero as $n \to \infty$ then we have a version of the central limit theorem for individual probabilities.

Theorem 4.3.3 Let Z_n be the normalised sum of n independent and identically distributed lattice variables with span h and variance σ^2. Then for each x specified by (4.3.4),

$$\lim_{n\to\infty} \frac{\sigma\sqrt{n}}{h} \Pr(Z_n = x) = \frac{1}{(2\pi)^{\frac{1}{2}}} e^{-\frac{1}{2}x^2}.$$

Proof. The second integral on the right hand side of (4.3.5) presents no difficulty since in absolute value it is dominated by

$$\frac{1}{2\pi} \int_{|s| > \pi\sigma\sqrt{n}/h} e^{-\frac{1}{2}s^2}\, ds = \frac{1}{(2\pi)^{\frac{1}{2}}} \left[1 - \Phi\left(\frac{\pi\sigma\sqrt{n}}{h}\right)\right]$$
$$\to 0 \quad \text{as } n \to \infty.$$

To show that the absolute value of the first integral on the right in (4.3.5) also tends to zero as $n \to \infty$ is difficult. That

$$\lim_{n\to\infty} |Ee^{isZ_n} - e^{-\frac{1}{2}s^2}| = 0 \tag{4.3.6}$$

is known from the proof for Theorem 4.3.1, but further argument is needed to obtain the result we want, namely

$$I_n = \int_{-\pi\sigma\sqrt{n}/h}^{\pi\sigma\sqrt{n}/h} |Ee^{isZ_n} - e^{-\frac{1}{2}s^2}|\, ds \to 0.$$

The procedure is to decompose I_n into the sum of three integrals, symbolically expressed by

$$\int_{-\pi\sigma\sqrt{n}/h}^{\pi\sigma\sqrt{n}/h} = \int_{|s| \leq a} + \int_{a < |s| \leq \delta\sigma\sqrt{n}} + \int_{\delta\sigma\sqrt{n} < |s| \leq \pi\sigma\sqrt{n}/h} \tag{4.3.7}$$

Here δ and a are to be chosen and it is required to show that each of these three integrals can be made arbitrarily small. The first one is easy since a consequence of (4.3.6) is

$$\lim_{n\to\infty} \int_{-a}^{a} |Ee^{isZ_n} - e^{-\frac{1}{2}s^2}|\, ds = 0$$

uniformly for every finite a.

Secondly, the Taylor expansion

$$Ee^{is(Y-\mu)} = 1-\tfrac{1}{2}s^s\sigma^2+o(s^2) \qquad \text{as } s \to 0$$

implies the inequality

$$|Ee^{is(Y-\mu)}| \leqslant 1-\tfrac{1}{4}s^2\sigma^2 \leqslant e^{-\frac{1}{4}s^2\sigma^2}$$

for s sufficiently small, say $|s| \leqslant \delta$. Then for $|s| \leqslant \delta\sigma\sqrt{n}$,

$$\begin{aligned} |Ee^{isZ_n}| &= \left|\left[E\exp\left\{\frac{is}{\sigma\sqrt{n}}(Y-\mu)\right\}\right]^n\right| \\ &\leqslant \left[\exp\left(-\frac{s^2\sigma^2}{4\sigma^2 n}\right)\right]^n \\ &= e^{-\frac{1}{4}s^2} \end{aligned}$$

and

$$|Ee^{isZ_n}-e^{-\frac{1}{2}s^2}| \leqslant 2e^{-\frac{1}{4}s^2} \quad \text{if } |s| \leqslant \delta\sigma\sqrt{n}.$$

With this choice of δ the middle integral on the right in (4.3.7) is

$$\begin{aligned} \int\limits_{a<|s|\leqslant\delta\sigma\sqrt{n}} |Ee^{isZ_n}-e^{-\frac{1}{2}s^2}|\,ds &\leqslant \int\limits_{a<|s|\leqslant\delta\sigma\sqrt{n}} 2e^{-\frac{1}{4}s^2}\,ds \\ &\leqslant 2e^{-\frac{1}{4}a^2}, \end{aligned}$$

which can be made arbitrarily small by making a sufficiently large.

Finally the extreme right-hand integral in (4.3.7) is dominated by

$$\int\limits_{\delta\sigma\sqrt{n}<|s|\leqslant\pi\sigma\sqrt{n}/h} |Ee^{isZ_n}|\,ds + \int\limits_{\delta\sigma\sqrt{n}<|s|\leqslant\pi\sigma\sqrt{n}/h} e^{-\frac{1}{2}s^2}\,ds.$$

The second of these clearly has the limit zero. Concerning the first recall from Exercise 2.3.17 that for any $\delta > 0$ it is possible if Y is lattice to find a positive number $\gamma = \gamma(\delta)$ such that

$$|Ee^{is(Y-\mu)}| \leqslant e^{-\gamma} \quad \text{if } \delta \leqslant |s| < \pi/h,$$

Then

$$\begin{aligned} |Ee^{isZ_n}| &= |(Ee^{is(Y-\mu)/\sigma\sqrt{n}}|^n \\ &\leqslant e^{-n\gamma} \quad \text{if } \delta\sigma\sqrt{n} \leqslant |s| < \pi\sigma\sqrt{n}/h \end{aligned}$$

and therefore

$$\int_{\delta\sigma\sqrt{n}<|s|\leqslant\pi\sigma\sqrt{n}/h} |Ee^{isZ_n}|\, ds \leqslant 2\sigma\sqrt{n}e^{-n\gamma}(\pi/h-\delta)$$

$$\to 0 \quad \text{as } n \to \infty$$

The proof is now complete. □

For example, if S_n is binomial with parameter p then $a = 0, h = 1$, $\mu = p, \sigma = (p(1-p))^{\frac{1}{2}}$, and as $n \to \infty$

$$\Pr(S_n = k) \sim \frac{1}{(2\pi np(1-p))^{\frac{1}{2}}} \exp\left[-\frac{(k-np)^2}{2np(1-p)}\right] \qquad k = 0, 1, 2\ldots$$

Exercises

1. A die is rolled 600 times. Find the probability that the number of 'sixes' obtained lies between 90 and 110. What is this probability for the number of 'threes'?

2. Let Y denote the average of the values shown when k dice are rolled. Find the probability that the sum of 600 such averages lies in the interval $100 \pm 90k^{-\frac{1}{2}}$.

3. Suppose $X(t)$ is the position at time t of a particle on the line subject to numerous infinitesimal displacements, each of which is positive or negative with equal probability. It is reasonable to suppose that the central limit theorem applies and that $X(t)$ is approximately normally distributed. Supposing the particle is initially at the origin, let $f(x; t)\, dx$ be the probability of finding $X(t)$ in $(x, x + dx)$ after time t. According to physical theory $f(x; t)$ satisfies the *diffusion equation*

$$\frac{\partial f(x \cdot t)}{\partial t} = \frac{\delta}{2} \frac{\partial^2 f(x; t)}{\partial x^2}.$$

Show that a solution is in fact

$$f(x; t) = \frac{e^{-x^2/2\delta t}}{(2\pi\delta t)^{\frac{1}{2}}}.$$

[*Ref*: FELLER I, pages 354–9. For an interesting general discussion see the article 'Brownian motion and potential theory' by R. HERSH and R. J. GRIEGO, pages 66–74 of *Scientific American*, March 1969.]

4. Apply the central limit theorem to a sequence of Poisson variables with parameter 1 to prove

$$\lim_{n\to\infty} e^{-n} \sum_{j=0}^{n} \frac{n^j}{j!} = \frac{1}{2}.$$

5. Compute the bound in the Berry–Esséen theorem when $n = 49$ and

(a) $\Pr(Y = 1) = p, \quad \Pr(Y = 0) = 1-p,$
(b) Y is negative exponential.

6. *The empirical distribution function.* $X_1, X_2, \ldots, X_n$, are independent random variables with distribution function $F(x)$. Let $I_j(x), j = 1, 2, \ldots, n$, be the indicator functions $I_j(x) = 1$ if $X_j \leqslant x$, $I_j(x) = 0$ otherwise. Their average $n^{-1}S_n(x) = n^{-1} \sum_{j=1}^{n} I_j(x)$ is called the empirical distribution function. For each fixed x show that $n^{-1}S_n(x) \xrightarrow{p} F(x)$ and that $\{I_n(x)\}$ obeys the central limit theorem.

7. (X, Y) are jointly distributed with zero means, variances σ_1^2 and σ_2^2 respectively, correlation coefficient ρ, and $(X_j, Y_j), j = 1, 2, \ldots, n$, are independently and identically distributed as (X, Y). Use the characteristic function $Ee^{isX+itY}$ and the continuity theorem to prove the bivariate central limit theorem

$$\lim_{n\to\infty} \Pr\left(\frac{1}{\sigma_1\sqrt{n}} \sum_{j=1}^{n} X_j \leqslant x, \frac{1}{\sigma_2\sqrt{n}} \sum_{j=1}^{n} Y_j \leqslant y\right)$$

$$= \frac{1}{2\pi(1-\rho^2)^{\frac{1}{2}}} \int_{-\infty}^{x} \int_{-\infty}^{y} \exp\left[-\frac{(x_1^2 - 2\rho x_1 y_1 + y_1^2)}{2(1-\rho^2)}\right] dx_1 dy_1.$$

8. *The empirical characteristic function.* A result similar to that of Exercise 6 holds also for characteristic functions. Let the real and imaginary parts of Ee^{isX} be respectively

$$U(s) = \int_{-\infty}^{\infty} f(x) \cos sx \, dx, \qquad V(s) = \int_{-\infty}^{\infty} f(x) \sin sx \, dx.$$

The corresponding empirical quantities are

$$U_n(s) = n^{-1} \sum_{j=1}^{n} \cos sX_j, \qquad V_n(s) = n^{-1} \sum_{j=1}^{n} \sin sX_j$$

Use the result of Exercise 7 to show that for each fixed $s \neq 0$,

$$\Pr\left(\frac{\{U_n(s)-U(s)\}(2n)^{\frac{1}{2}}}{[1+U(2s)-2\{U(s)\}^2]^{\frac{1}{2}}} \leqslant x, \frac{\{V_n(s)-V(s)\}(2n)^{\frac{1}{2}}}{[1-U(2s)-2\{V(s)\}^2]^{\frac{1}{2}}} \leqslant y\right)$$

has the bivariate normal distribution function for a limit. Find ρ.

9. (Continuation.) If X is symmetric about the origin, then

$$V_n(s)(2n)^{\frac{1}{2}}/(1-U(2s))^{\frac{1}{2}}$$

converges in distribution to the standard normal.

10. *Lindeberg's Theorem* states: Suppose $X_1, X_2, \ldots, X_n$, are independent random variables with

$$EX_j = m_j, \quad \operatorname{Var} X_j = \sigma_j^2, \quad j = 1, 2, \ldots, n,$$

and let $S_n = X_1 + X_2 + \ldots + X_n$. Then $\{X_n\}$ obeys the central limit theorem if for every $\varepsilon > 0$

$$\lim_{n\to\infty} \frac{1}{\operatorname{Var} S_n} \sum_{k=1}^{n} \int_{|x-m_k|>(\operatorname{Var} S_n)^{\frac{1}{2}}} (x-m_k)^2 f_k(x)\, dx = 0.$$

We have written $f_k(x)$ as the density of X_k. In the discrete case the integral must be replaced by a sum. Feller subsequently proved the necessity of this condition provided also that as $n \to \infty$.

$$\operatorname{Var} S_n \to \infty \quad \text{and} \quad \sigma_n^2/\operatorname{Var} S_n \to 0.$$

Which of the following sequences satisfy the conditions of the theorem

(a) $\Pr(X_j = \pm 2^j) = \frac{1}{2}$,
(b) $\Pr(X_j = \pm a^j) = 1/a^{2j}$, $\Pr(X_j = 0) = 1 - 2a^{-2j}$, $a > 1$,
(c) $\Pr(X_j = \pm j^a) = \frac{1}{2}$.

4.4* DISTRIBUTIONS DERIVED FROM THE NORMAL

Mainly because of the central limit theorem, the norml distribution occupies a special place in probability theory and statistics. In this

* To be omitted at a first reading.

section we examine some special distributions that can be derived from the normal and play a useful role in statistical inference.

Let $X_1, X_2, \ldots, X_n$ be independent and identically distributed as the $N(\mu, \sigma^2)$ variable X. That is, their common density function is

$$f(x) = \frac{1}{\sigma(2\pi)^{\frac{1}{2}}} \exp\left[-\frac{1}{2}\left(\frac{x-\mu}{\sigma}\right)^2\right], \qquad -\infty < x < \infty,$$

with joint density

$$f(x_1, x_2, \ldots, x_n) = \frac{1}{\sigma^n(2\pi)^{\frac{1}{2}n}} \exp\left\{-\frac{1}{2\sigma^2}\sum_{j=1}^{n}(x_j-\mu)^2\right\}$$

and moment generating function

$$E\exp(-\sum_{j=1}^{n} s_j X_j) = \prod_{j=1}^{n} \exp(-\mu s_j + \tfrac{1}{2}\sigma^2 s_j^2).$$

$$= \exp(-\mu \sum_{j=1}^{n} s_j + \tfrac{1}{2}\sigma^2 \sum_{j=1}^{n} s_j^2). \qquad (4.4.2)$$

For *fixed n* it will be convenient to refer to these $X_1, \ldots, X_n$ as an *independent sample* of n from a $N(\mu, \sigma^2)$ population. A realisation of the sample $(x_1, x_2, \ldots, x_n)$ corresponds to n independent realisations of the $N(\mu, \sigma^2)$ random variable X. The *sample average* or *sample mean* is

$$\overline{X} = \frac{1}{n}(X_1 + X_2 + \ldots + X_n)$$

with expectation

$$E\overline{X} = \frac{1}{n} n\mu = \mu. \qquad (4.4.3)$$

If the value of μ is unknown and one uses realisations of the sample average $\overline{X}$ to estimate it then (4.4.3) implies that $\overline{X}$ is what is called an *unbiased estimator* of μ. From (4.4.2) we can obtain the way in which $\overline{X}$ is distributed about its expectation μ, since writing $s_j = s$,

$$E\exp(-s\overline{X}) = E\exp\left[-\frac{s}{n}(X_1, + X_2 + \ldots + X_n)\right]$$

$$= \exp\left[-\mu s + \frac{\sigma^2 s^2}{2n}\right].$$

Hence $\bar{X}$ is also normally distributed with mean μ and variance σ^2/n.

Example 1 Suppose $X \frown N(\mu, \sigma_0^2)$ where μ is unknown but the variance $\sigma^2 = \sigma_0^2$ is given. Then

$$\begin{aligned}\Phi(b) - \Phi(a) &= \Pr\left(a < \frac{(\bar{X}-\mu)\sqrt{n}}{\sigma_0} < b\right) \\ &= \Pr\left(\bar{X} - \frac{b\sigma_0}{\sqrt{n}} < \mu < \bar{X} - \frac{a\sigma_0}{\sqrt{n}}\right)\end{aligned}$$

yields an interval estimate of the unknown mean μ. In particular if $b = 1{\cdot}96$, $a = -1{\cdot}96$, then from tables

$$\Pr\left(\bar{X} - \frac{1{\cdot}96\sigma_0}{\sqrt{n}} < \mu < \bar{X} + \frac{1{\cdot}96\sigma_0}{\sqrt{n}}\right) = 0{\cdot}95.$$

If $\bar{x}$ is a realised value of $\bar{X}$ the interval

$$(\bar{x} - 1{\cdot}96\sigma_0/\sqrt{n}, \quad \bar{x} + 1{\cdot}96\sigma_0/\sqrt{n})$$

is called a 95% *confidence interval* for μ. □

We introduce now the *sample variance*

$$S^2 = \frac{1}{n-1} \sum_{j=1}^{n} (X_j - \bar{X})^2, \qquad n \geqslant 2, \tag{4.4.4}$$

which is not defined if $n = 1$. The reason for dividing the sum of squares $\sum_{j=1}^{n} (X_j - \bar{X})^2$ by $n-1$ instead of n is that

$$ES^2 = \sigma^2. \tag{4.4.5}$$

That is, the sample variance defined by (4.4.4) is an unbiased estimator of the population variance σ^2. To see this write

$$\begin{aligned}\sum_{j=1}^{n} (X_j - \bar{X})^2 &= \sum_{j=1}^{n} \{(X_j - \mu) - (\bar{X} - \mu)\}^2 \\ &= \sum_{j=1}^{n} (X_j - \mu)^2 - n(\bar{X} - \mu)^2.\end{aligned} \tag{4.4.6}$$

Since $X_j \frown N(\mu, \sigma^2)$ and $\bar{X} \frown N(\mu, \sigma^2/n)$ it follows by definition that

$$E(X_j - \mu)^2 = \sigma^2 \text{ and } E(\bar{X} - \mu)^2 = \sigma^2/n.$$

Therefore

$$ES^2 = \frac{1}{n-1} E \sum_{j=1}^{n} (X_j - \bar{X})^2$$

$$= \frac{1}{n-1}\left\{\sum_{j=1}^{n} E(X_j-\mu)^2 - nE(\bar{X}-\mu)^2\right\}$$

$$= \frac{1}{n-1}\left[n\sigma^2 - \frac{n\sigma^2}{n}\right] = \sigma^2,$$

as required.

The distribution of S^2 is more difficult to obtain and we will first require some facts about the chi-square distribution. Recall from Example 2.4.3 and Exercise 2.4.2 that if $Y \frown N(0,1)$ then $Z = Y^2$ is distributed as chi square with one degree of freedom, i.e. $Z \frown \chi^2_{(1)}$. The density function of a $\chi^2_{(1)}$ random variable is given by (2.4.8) and (2.5.6) and its moment generating function (Exercise 2.4.2) is

$$Ee^{-sZ} = (1+2s)^{-\frac{1}{2}}.$$

If $Z_1, Z_2, \ldots, Z_n$ are independent $\chi^2_{(1)}$ variables their sum

$$W_n = Z_1 + Z_2 + \ldots + Z_n$$

has moment generating function

$$Ee^{-sW_n} = \prod_{j=1}^{n} Ee^{-sZ_j} = (1+2s)^{-\frac{1}{2}n}. \tag{4.4.7}$$

But from the table on page 108 this is the generating function of the gamma density

$$f(x) = \frac{e^{-\frac{1}{2}x} x^{\frac{1}{2}n-1}}{\Gamma(\frac{1}{2}n) 2^{\frac{1}{2}n}}, \qquad 0 < x,$$

which is also called the chi-square density with n degrees of freedom (§2.5(i)).

Our result concerning the chi-square distribution is then that the sum of n independent $\chi^2_{(1)}$ variables is $\chi^2_{(n)}$. The more general result, apparent from (4.4.7), is that if $Z_1, Z_2, \ldots, Z_k$ are independent with $Z_j \frown \chi^2_{(m_j)}$, $j = 1, 2, \ldots, k$, then $Z_1 + Z_2 + \ldots + Z_k$ is distributed as chi square with $\sum_{j=1}^{k} m_j$ degrees of freedom. (c.f. Exercise 2.5.1).

o

This is relevant to the distribution of S^2 in the following way. After dividing through by σ^2 equation (4.4.6) can be rewritten as

$$\sum_{j=1}^{n}\left[\frac{X_j-\mu}{\sigma}\right]^2 = \frac{(n-1)S^2}{\sigma^2}+\left[\frac{\overline{X}-\mu}{\sigma/\sqrt{n}}\right]^2. \tag{4.4.8}$$

Since $(X_j - \mu)/\sigma \frown N(0, 1)$ it follows that the sum on the left-hand side of (4.4.8) is $\chi^2_{(n)}$. Also, $(\overline{X} - \mu)\sqrt{n}/\sigma \frown N(0, 1)$ implies that

$$\left(\frac{\overline{X}-\mu}{\sigma/\sqrt{n}}\right)^2 \frown \chi^2_{(1)}.$$

If it were known that S^2 and $\overline{X}$ were independently distributed it would follow immediately that

$$Ee^{-s(n-1)S^2/\sigma^2} = \frac{(1+2s)^{\frac{1}{2}}}{(1+2s)^{\frac{1}{2}n}} = \frac{1}{(1+2s)^{\frac{1}{2}(n-1)}},$$

i.e. that $(n-1)S^2/\sigma^2$ is distributed as $\chi^2_{(n-1)}$.

It is a fact that S^2 and $\overline{X}$ are independent, but unfortunately the proof requires multidimensional change of variable techniques not dealt with here. For that reason we are forced to assume the independence of S^2 and $\overline{X}$. Under this assumption we can sum up our discussion as follows:

Theorem 4.4.1 Suppose $X_1, X_2, \ldots, X_n$ are independent $N(\mu, \sigma^2)$ variables and let

$$\overline{X} = \frac{1}{n}(X_1+X_2+\ldots+X_n), \qquad S^2 = \frac{1}{n-1}\sum_{j=1}^{n}(X_j-\overline{X})^2.$$

Then

$$\Pr(\overline{X} \leqslant x) = \frac{\sqrt{n}}{\sigma(2\pi)^{\frac{1}{2}}}\int_{-\infty}^{x} e^{-n(y-\mu)^2/2\sigma^2}\,dy,$$

and

$$\Pr\left(\frac{(n-1)S^2}{\sigma^2} \leqslant x\right) = \int_0^x \frac{e^{-(y/2)}y^{\frac{1}{2}(n-1)-1}}{2^{\frac{1}{2}(n-1)}\Gamma(\frac{1}{2}(n-1))}\,dy. \qquad \square$$

The independence of sample mean and variance is an important property of samples from a normal population. We require this result again when considering the sample analogue of the standardised variable

$$Z = \frac{(\bar{X} - \mu)\sqrt{n}}{\sigma},$$

namely

$$T = \frac{(\bar{X} - \mu)\sqrt{n}}{S}. \tag{4.4.9}$$

The random variable T can also be written

$$T = \left[\frac{\bar{X} - \mu}{\sigma/\sqrt{n}}\right] \bigg/ \left[\frac{(n-1)S^2}{(n-1)\sigma^2}\right]^{\frac{1}{2}}, \tag{4.4.10}$$

or more suggestively as

$$T = \frac{N(0, 1)}{\left[\frac{1}{n-1}\chi^2_{(n-1)}\right]^{\frac{1}{2}}},$$

the ratio of a standard normal variable to the square root of a chi-square variable divided by its degrees of freedom. The following lemma gives a method of finding the density of a ratio.

Lemma: If Y_1 and Y_2 are independent and $Y_2 > 0$, the density of $Y_3 = Y_1/Y_2$ is

$$f_{Y_3}(x) = \int_0^\infty y f_{Y_1}(xy) f_{Y_2}(y)\, dy. \tag{4.4.11}$$

Proof

$$\begin{aligned}
\Pr(Y_3 \leqslant x) &= \Pr(Y_1 \leqslant xY_2) \\
&= \int_0^\infty \Pr(Y_1 \leqslant xy \mid Y_2 = y) f_{Y_2}(y)\, dy \\
&= \int_0^\infty \Pr(Y_1 \leqslant xy) f_{Y_2}(y)\, dy, \quad \text{by independence.}
\end{aligned}$$

The density $f_{Y_3}(x)$ of Y_3 is found by differentiation to be as specified in (4.4.11). □

The density of T defined in (4.4.10) is an immediate consequence of the following more general result.

Theorem 4.4.2 Suppose Z and Y are independent and $Z \frown N(0,1)$, $Y \frown \chi^2_{(\nu)}$, and let

$$T = \frac{Z}{(Y/\nu)^{\frac{1}{2}}}.$$

Then the density function of T is

$$f_T(t) = \frac{1}{B(\frac{1}{2}\nu, \frac{1}{2})} \cdot \frac{1}{\nu^{\frac{1}{2}}(1+t^2/\nu)^{\frac{1}{2}(\nu+1)}}, \qquad -\infty < t < \infty. \tag{4.4.12}$$

In particular the density of the ratio in (4.4.10) is given by (4.4.12) with $\nu = n-1$.

Proof. Given $Y \frown \chi^2_{(\nu)}$, in Exercise 1 you are asked to show that $Y_2 = (Y/\nu)^{\frac{1}{2}}$ has density

$$f_{Y_2}(y) = \frac{e^{-\frac{1}{2}\nu y^2} y^{\nu-1} \nu^{\frac{1}{2}\nu}}{2^{\frac{1}{2}\nu-1}\Gamma(\frac{1}{2}\nu)}, \qquad 0 < y.$$

For this Y_2 and $Y_1 = Z \frown N(0,1)$ the density of $T = Z(Y/\nu)^{-\frac{1}{2}}$ is

$$f_T(t) = \int_0^\infty \frac{e^{-\frac{1}{2}(ty)^2}}{(2\pi)^{\frac{1}{2}}} \cdot \frac{e^{-\frac{1}{2}\nu y^2} y^{\nu-1} \nu^{\frac{1}{2}\nu} y}{2^{\frac{1}{2}\nu-1}\Gamma(\frac{1}{2}\nu)}\, dy$$

$$= \int_0^\infty \frac{e^{-\frac{1}{2}y^2(\nu+t^2)}(y\sqrt{\nu})^\nu}{2^{\frac{1}{2}\nu}\Gamma(\frac{1}{2})\Gamma(\frac{1}{2}\nu)}\, dy \qquad \text{since by (2.5.4) } \Gamma(\tfrac{1}{2}) = \sqrt{\pi}.$$

Substituting $x = \frac{1}{2}y^2(\nu+t^2)$,

$$f_T(t) = \frac{\int_0^\infty e^{-x} x^{\frac{1}{2}(\nu-1)}\, dx}{\Gamma(\frac{1}{2})\Gamma(\frac{1}{2}\nu)(1+t^2/\nu)^{\frac{1}{2}(\nu+1)}\sqrt{\nu}}$$

$$= \frac{\Gamma(\frac{1}{2}(\nu+1))}{\Gamma(\frac{1}{2})\Gamma(\frac{1}{2}\nu)} \cdot \frac{1}{(1+t^2/\nu)^{\frac{1}{2}(\nu+1)}\sqrt{\nu}}$$

The final equality and hence the theorem follow from the properties (2.5.1) and (2.5.8) of the gamma and beta functions. □

A random variable with the density (4.4.12) is said to be distributed* as *Student's t with v degrees of freedom*, and we use the notation

$$T \frown t_{(v)}.$$

Tables of the t distribution function are available and one application is to obtain interval estimates for μ (as in Example 5) when the variance σ^2 is unknown.

Finally we derive a distribution which arises when considering the ratio of two independent chi-square variables. This is the so called F distribution, mentioned on page 101 as a special case of the beta distribution. It has density function (c.f. 2.5.10)

$$f_F(x) = \frac{(m/n)^{\frac{1}{2}m} x^{\frac{1}{2}m-1}}{B(\frac{1}{2}m, \frac{1}{2}n)(1+mx/n)^{\frac{1}{2}(m+n)}}, \qquad 0 < x. \qquad (4.4.13)$$

A random variable with this density is said to satisfy the F distribution with (m, n) degrees of freedom, abbreviated $F_{(m, n)}$.

Theorem 4.4.3 If Y and Z are independent and $Y \frown \chi^2_{(m)}$, $Z \frown \chi^2_{(n)}$, then

$$F = \frac{Y/m}{Z/n}$$

is distributed as $F_{(m, n)}$ with density (4.4.13).

Proof. If $Y \frown \chi^2_{(m)}$ the density of Y/m is

$$f_{Y/m}(y) = \frac{e^{-\frac{1}{2}my} y^{\frac{1}{2}m-1} m^{\frac{1}{2}m}}{2^{\frac{1}{2}m}\Gamma(\frac{1}{2}m)}$$

Then from (4.4.11) the density of $F = \dfrac{Y/m}{Z/n}$ is

$$f_F(x) = \int_0^\infty \frac{e^{-\frac{1}{2}mxy}(xy)^{\frac{1}{2}m-1} m^{\frac{1}{2}m} y e^{-\frac{1}{2}ny} y^{\frac{1}{2}n-1} n^{\frac{1}{2}n}}{2^{\frac{1}{2}m}\Gamma(\frac{1}{2}m) . 2^{\frac{1}{2}n}\Gamma(\frac{1}{2}n)} \, dy$$

$$= \frac{m^{\frac{1}{2}m} n^{\frac{1}{2}n} x^{\frac{1}{2}m-1}}{2^{\frac{1}{2}(m+n)}\Gamma(\frac{1}{2}m)\Gamma(\frac{1}{2}n)} \int_0^\infty e^{-\frac{1}{2}y(n+mx)} y^{\frac{1}{2}(m+n)-1} \, dy$$

* After W. S. Gosset who wrote under the pseudonym 'Student'.

Substituting $z = \frac{1}{2}y(n+mx)$,

$$f_F(x) = \frac{m^{\frac{1}{2}m}x^{\frac{1}{2}m-1}}{\Gamma(\frac{1}{2}m)\Gamma(\frac{1}{2}n)n^{\frac{1}{2}m}(1+mx/n)^{\frac{1}{2}(m+n)}} \int_0^\infty e^{-z} z^{\frac{1}{2}(m+n)-1}\, dz$$

which, on referring to the definition of the gamma and beta functions, is seen to be identical to (4.4.13). □

Recall that $(\overline{X}-\mu)\sqrt{n}/\sigma \frown N(0,1)$ and hence $(\overline{X}-\mu)^2 n/\sigma^2 \frown \chi^2_{(1)}$. Therefore the square of the random variable T in (4.4.10),

$$T^2 = \frac{(\overline{X}-\mu)^2 n}{S^2},$$

is distributed according to $F_{(1,n-1)}$. This fact can be used for an alternative derivation of the $t_{(n-1)}$ density.

It should be emphasised that the results given in this section were obtained under the assumption that the variables $X_1, X_2, \ldots, X_n$ were independent and normally distributed. The proofs are not valid when this assumption is violated, but nonetheless the results are widely used in statistical inference as approximations even when the underlying population is not normal.

Exercises

1. Suppose $Y \frown \chi^2_{(v)}$. Complete the proof of Theorem 4.4.2 by finding the density of $Y_2 = \sqrt{(Y/v)}$.

2. X_1 and X_2 are independent $N(0,1)$ variables. Show that $Z = X_1/X_2$ is Cauchy.

3. Find the density of $Z = X_1/X_2$ if X_1 and X_2 are independent with common density $f(x) = e^{-x}, 0 < x$.

4. Show that if $X \frown F_{(m,n)}$ then $Y = 1/X \frown F_{(n,m)}$.

5. $\overline{X}$ and S^2 are the mean and variance of a sample of n from a $N(\mu, \sigma^2)$ population. If $T \frown t_{(n-1)}$ and for a given α the number b (obtained from tables) is such that $\Pr(-b \leqslant T \leqslant b) = 1-\alpha$, show that a $100(1-\alpha)\%$ confidence interval for μ is $(\overline{x}-bs/\sqrt{n}, \overline{x}+bs/\sqrt{n})$.

6. (Continuation.) Suppose $n = 6, \bar{x} = 4, s^2 = 9$. In the light of this evidence is the assumption that $\mu = 6$ reasonable (take $\alpha = 0{\cdot}05$)? Is your conclusion the same if in fact $n = 30$?

7. If the mean μ is known it is natural to define the sample variance as

$$S_1^2 = n^{-1} \sum_{j=1}^{n} (X_j - \mu)^2.$$

Obtain a $100(1-\alpha)\%$ confidence interval for the variance σ^2 when
(a) μ is known
(b) μ is estimated by $\bar{X}$.

8. X and Y are normal variables with means μ_1, μ_2 respectively and samples of n X observations and m Y observations are taken. Assuming that Var X = Var $Y = \sigma^2$ find a $100(1-\alpha)\%$ confidence interval for $\mu_1 - \mu_2$ when
(a) σ^2 is known
(b) σ^2 must be estimated from the samples.

9. (Continuation.) If $n = 16, m = 9, \bar{x} = 2, \bar{y} = 3$, is the assertion $\mu_1 = \mu_2$ plausible if the common variance is known to be

(a) $\sigma^2 = 1$ (b) $\sigma^2 = 25$?

10. Show how the F distribution can be used to check the assumption that the variances of two normal populations are identical.

4.5* SOME LIMIT THEOREMS FOR MARKOV CHAINS

The major results of this chapter have been limit theorems for the distribution of the partial sum S_n of independent and identically distributed random variables. For integer valued summands it was shown in Example 3.5.2 that such a sequence $\{S_n\}$ is a discrete parameter homogeneous Markov chain on the integers with transition probabilities $p_{ij}^{(n)} = \Pr(S_n = j \mid S_0 = i)$. Some information about the limit behaviour of these $p_{ij}^{(n)}$ is contained in the law of large

* To be omitted at a first reading.

numbers and the central limit theorem. For example the latter result translates in Markov chain notation as

$$\lim_{n\to\infty} \sum_{\substack{a\sigma\sqrt{n}\leqslant j-n\mu \\ \leqslant b\sigma\sqrt{n}}} p_{oj}^{(n)} = \frac{1}{(2\pi)^{\frac{1}{2}}}\int_a^b e^{-\frac{1}{2}x^2}\,dx.$$

In this section we consider the question; given a discrete parameter homogeneous Markov chain $\{X_n\}$, what can be said about the behaviour of the $p_{ij}^{(n)}$ as $n \to \infty$? In particular, if the limits exist do they constitute a probability distribution? If this turns out to be the case we have that $\{X_n\}$ is convergent in distribution.

If $\mathscr{P} = (p_{ij})$ is the stochastic matrix of one-step transition probabilities, recall from §3.5 that $\mathscr{P}^n$ is the matrix whose (i, j)th term is $p_{ij}^{(n)}$. Seeking the limits $\lim_{n\to\infty} p_{ij}^{(n)}$ for all states is then equivalent to asking for $\lim_{n\to\infty} \mathscr{P}^n$. If the state space consists of only a finite number of states, so that $\mathscr{P}$ is a finite matrix, the special properties of finite non-negative matrices can be used to solve the limit problem. These methods fail if $\mathscr{P}$ is infinite, and a more general approach is required. The aim of this section is to present a brief outline of this general theory, applicable to both finite and infinite chains.

The methods used earlier in this chapter relied heavily on the fact that the generating function of the distribution of the sum of n independent and identically distributed variables is the nth power of a generating function. This is of course not true of Markov chains generally. Nonetheless generating functions continue to play an important role but for a different reason.

For any state i the transition probability

$$p_{ii}^{(n)} = \Pr(X_n = i \mid X_0 = i)$$

is the probability that a return to i occurs at time n. This must be distinguished from the probability that the *first* return to i occurs at time n, namely

$$f_{ii}^{(n)} = \Pr(X_n = i;\, X_v \neq i,\, v = 1, 2, \ldots, n-1 \mid X_0 = i),\ n = 2, 3, \ldots, \tag{4.5.1}$$

$$f_{ii}^{(0)} = 0, \quad f_{ii}^{(1)} = p_{ii}.$$

When $i \neq j$ (4.5.1) is replaced by

$$f_{ij}^{(n)} = \Pr(X_n = j;\, X_\nu \neq j,\, \nu = 1,2,\ldots,n-1 \,|\, X_0 = i), \qquad n = 2,3,4,\ldots,$$
$$f_{ij}^{(0)} = 0, \quad f_{ij}^{(1)} = p_{ij}. \tag{4.5.2}$$

For fixed i and j $f_{ij}^{(n)}$ is the probability that *first passage* from i to j occurs at time n.

For all pairs of states (i,j) the sequence $\{f_{ij}^{(n)}, n = 1,2,\ldots\}$ is a (possibly dishonest) probability distribution with

$$F_{ij} = \sum_{n=1}^{\infty} f_{ij}^{(n)} \leqslant 1.$$

The sum F_{ij} is the probability that state j is ever visited from i, and the defect $1-F_{ij}$ (which of course may be zero) is the probability of the complementary event that j is never reached from i. When $F_{ij} = 1$ for given i and j $\{f_{ii}^{(n)}, n = 1,2,\ldots\}$ is called the *recurrence time* distribution of state i in distinction to the term *first passage distribution* applied to $\{f_{ij}^{(n)}\}$ when $i \neq j$.

It is important to note that $\{f_{ij}^{(n)}, n = 1,2,\ldots\}$ is a distribution over time whereas the set of transition probabilities

$$\{p_{ij}^{(n)}, j = 0, \pm 1, \ldots\}$$

is a distribution over the state space. A relationship of fundamental importance holds between the $f_{ij}^{(n)}$ and $p_{ij}^{(n)}$. Suppose first that $i = j$.

Keeping i fixed note that the event 'return to i at time n' can be decomposed into the union of the disjoint events 'return to i for the first time at time ν, and return again $n-\nu$ time units later'. Thus for $n = 2,3,\ldots$

$$\begin{aligned} p_{ii}^{(n)} &= \sum_{\nu=1}^{n} \Pr(X_\nu = i;\, X_m \neq i,\, m = 1,2,\ldots,\nu-1;\, X_n = i \,|\, X_0 = i) \\ &= \sum_{\nu=1}^{n} \Pr(X_\nu = i;\, X_m \neq i,\, m = 1,2,\ldots,\nu-1 \,|\, X_0 = i) \\ &\quad \Pr(X_n = i \,|\, X_\nu = i;\, X_m \neq i, m = 1,2,\ldots,\nu-1, X_0 = i)) \\ &= \sum_{\nu=1}^{n} f_{ii}^{(\nu)} p_{ii}^{(n-\nu)}, \end{aligned} \tag{4.5.3}$$

where the Markov property has been used to obtain

$$\begin{aligned}\Pr(X_n = i \mid X_\nu = i;\, X_m \neq i, m = 1,2,\dots,\nu-1, X_0 = i) &\\ = \Pr(X_n = i \mid X_\nu = i) &\\ = \Pr(X_{n-\nu} = i \mid X_0 = i) &\\ = p_{ii}^{(n-\nu)}. &\end{aligned}$$

When $n = 0$ or 1 we have

$$p_{ii}^{(0)} = 1, \qquad p_{ii}^{(1)} = p_{ii} = f_{ii}^{(1)}. \tag{4.5.4}$$

Recognising that the right hand side of (4.5.3) is a convolution, it is natural to introduce the generating functions

$$G_{ii}(z) = \sum_{n=0}^{\infty} z^n p_{ii}^{(n)}, \qquad F_{ii}(z) = \sum_{n=1}^{\infty} z^n f_{ii}^{(n)}.$$

Multiplying (4.5.3) and (4.5.4) by the appropriate power z^n and summing over n yields

$$\sum_{n=0}^{\infty} z^n p_{ii}^{(n)} = 1 + \sum_{n=1}^{\infty} z^n \sum_{\nu=1}^{n} f_{ii}^{(\nu)} p_{ii}^{(n-\nu)}.$$

That is, from the convolution Theorem 2.2.4,

$$G_{ii}(z) = 1 + G_{ii}(z) F_{ii}(z),$$

or

$$G_{ii}(z) = \frac{1}{1 - F_{ii}(z)}. \tag{4.5.5}$$

A corresponding relationship holds when $i \neq j$. By an argument similar to that leading to (4.5.3) we find for $i \neq j$

$$\left.\begin{aligned} p_{ij}^{(n)} &= \sum_{\nu=1}^{n} f_{ij}^{(\nu)} p_{jj}^{(n-\nu)}, \qquad n = 2,3,\dots, \\ p_{ij}^{(0)} &= 0, \qquad p_{ij}^{(1)} = f_{ij}^{(1)}. \end{aligned}\right. \tag{4.5.6}$$

For the generating functions

$$G_{ij}(z) = \sum_{n=0}^{\infty} z^n p_{ij}^{(n)}, \qquad F_{ij}(z) = \sum_{n=1}^{\infty} z^n f_{ij}^{(n)},$$

(4.5.6) implies

$$G_{ij}(z) = F_{ij}(z) G_{jj}(z) = \frac{F_{ij}(z)}{1 - F_{jj}(z)}, \qquad i \neq j. \tag{4.5.7}$$

Two separate cases arise if one attempts to calculate the limits of the transition probabilities from (4.5.5) and (4.5.7). Firstly suppose that

$$F_{ii} = \sum_{n=1}^{\infty} f_{ii}^{(n)} < 1,$$

i.e. suppose that the chain is not certain to return to the starting state. From Abel's Theorem 2.2.2

$$\lim_{z \to 1-} F_{ii}(z) = F_{ii} < 1$$

and hence, from (4.5.5),

$$\sum_{n=0}^{\infty} p_{ii}^{(n)} = \lim_{n \to 1-} G_{ii}(z) = \frac{1}{1-F_{ii}} < \infty.$$

Applying the same argument to (4.5.7) we see that

$$\sum_{n=0}^{\infty} p_{ij}^{(n)} = \frac{F_{ij}}{1-F_{jj}} < \infty$$

if $F_{jj} < 1$. Conversely, $F_{jj} < 1$ if $\sum_{n=0}^{\infty} p_{ij}^{(n)} < \infty$.

Definition 4.5.1 State j is said to be *transient* if $F_{jj} < 1$, or equivalently if $\sum_{n=0}^{\infty} p_{jj}^{(n)} < \infty$. □

An immediate consequence is

Theorem 4.5.1 If state j is transient then for all k $\lim_{n \to \infty} p_{kj}^{(n)} = 0$. □

The question is much more difficult when state j is not transient, that is when $F_{jj} = 1$.

Definition 4.5.2 State j is said to be *recurrent* if $F_{jj} = 1$, or equivalently if $\sum_{n=0}^{\infty} p_{jj}^{(n)} = \infty$. □

The possibility of a non-zero limit for $p_{jj}^{(n)}$ and $p_{kj}^{(n)}$ therefore arises only if j is recurrent for then, again by Abel's Theorem, we have

$$\sum_{n=0}^{\infty} p_{jj}^{(n)} = \lim_{n\to 1-} \frac{1}{1-F_{jj}(z)} = \infty.$$

Consider first (4.5.5). If we are prepared to *assume* that $\lim_{n\to\infty} p_{jj}^{(n)}$ exists we can calculate its value using (2.2.7) thus

$$\lim_{n\to\infty} p_{jj}^{(n)} = \lim_{z\to 1-} (1-z)G_{jj}(z)$$

$$= \lim_{z\to 1-} \frac{1-z}{1-F_{jj}(z)} = \frac{1}{F'_{jj}(1)}. \tag{4.5.8}$$

The proof that in fact $p_{jj}^{(n)}$ does converge to a limit is difficult* and we will merely assume the truth of this proposition.

At this stage we require the notion of periodicity. It may so happen that $p_{jj}^{(n)}$ is zero unless n is a multiple of some integer $d \geqslant 2$.

Definition 4.5.3 State j is said to be *periodic* with period d if $p_{jj}^{(n)} = 0$ unless n is a multiple of d and d is the largest integer with this property. If there is no $d \geqslant 2$ with this property then j is *aperiodic*. □

The reason for introducing this notion is that if j is periodic then only every dth term of the sequence $\{p_{jj}^{(n)}, n = 1, 2, \ldots\}$ may be non-zero, in which case the sequence is alternating and $\lim_{n\to\infty} p_{jj}^{(n)}$ does not exist. However, our theory applies if the limit is taken through the subsequence $n = d, 2d, 3d, \ldots$ and for periodic states our results must be modified accordingly. To avoid this nuisance we will assume in future that all states are aperiodic.

Returning to (4.5.8) recognise that $F'_{jj}(1) = \mu_{jj} = \sum_{n=1}^{\infty} nf_{jj}^{(n)}$ is the mean of the recurrence time distribution. Hence

Theorem 4.5.2 If j is an aperiodic recurrent state then

$$\lim_{n\to\infty} p_{jj}^{(n)} = 1/\mu_{jj}.$$ □

* See FELLER I, pages 335–338. This is the basic existence theorem for Markov chains but its proof is beyond the scope of our brief discussion.

The mean recurrence time μ_{jj} is the first moment of a distribution on the positive integers and it is therefore possible that $\mu_{jj} = \infty$. Hence $p_{jj}^{(n)}$ converges to a strictly positive limit if and only if μ_{jj} is finite.

Definition 4.5.4 Suppose j is aperiodic and recurrent. Then j is said to be

(i) *null recurrent* if $\mu_{jj} = \infty$, in which case $p_{jj}^{(n)} \to 0$,
(ii) *positive recurrent* if $\mu_{jj} < \infty$, in which case

$$\lim_{n\to\infty} p_{jj}^{(n)} = 1/\mu_{jj} > 0. \qquad \square$$

Similar arguments applied to (4.5.7) yield the corresponding result for $p_{ij}^{(n)}$ when $i \neq j$:

Theorem 4.5.3 Let j be aperiodic and recurrent. If j is

(i) null recurrent, then $\lim_{n\to\infty} p_{ij}^{(n)} = 0$,
(ii) positive recurrent, then $\lim_{n\to\infty} p_{ij}^{(n)} = F_{ij}/\mu_{jj}$. $\square$

We require one more concept. In the derivation of (4.5.6) it was tacitly assumed that j could be reached from i, namely that there exists a time $n > 0$ such that $p_{ij}^{(n)} > 0$, for otherwise (4.5.6) is vacuous.

Definition 4.5.5 A Markov chain with the property that to *every* pair of states (i, j) there exists $n > 0$ such that $p_{ij}^{(n)} > 0$ is said to be *irreducible*. $\square$

Irreducible chains are then those in which any state can be visited from any other. Our discussion will be confined to irreducible chains (Exercises 10 and 11 treat examples of the contrary case).

At this point it is convenient to formally restate an identity you were asked to verify in Exercise 3.5.1:

Lemma. For every pair of states (i, j) the transition probabilities $p_{ij}^{(n)}$ of all discrete parameter Markov chains satisfy the Chapman–Kolmogorov equations

$$p_{ij}^{(n+m)} = \sum_k p_{ik}^{(n)} p_{kj}^{(m)}. \qquad (4.5.9) \ \square$$

Theorem 4.5.4 In an irreducible chain all states are of the same type. That is, they all have the same period d or are all aperiodic, and are all transient, null recurrent, or positive recurrent together.

Proof. Let i and j be any two given states. By irreducibility there exist $m > 0$ and $r > 0$ such that $p_{ij}^{(m)} = a > 0$ and $p_{ji}^{(r)} = b > 0$. From (4.5.9), for any $n > 0$ and these m, r,

$$\begin{aligned} p_{ii}^{(m+n+r)} &= \sum_k p_{ik}^{(m+n)} p_{ki}^{(r)} \\ &\geqslant p_{ij}^{(m+n)} p_{ji}^{(r)} \\ &= p_{ji}^{(r)} \sum_k p_{ik}^{(m)} p_{kj}^{(n)} \\ &\geqslant p_{ji}^{(r)} p_{ij}^{(m)} p_{jj}^{(n)} \\ &= ab p_{jj}^{(n)}. \end{aligned} \tag{4.5.10}$$

Reversing the roles of i and j we also have

$$p_{jj}^{(m+n+r)} \geqslant ab p_{ii}^{(n)} \tag{4.5.11}$$

The theorem is a consequence of these two inequalities. Consider periodicity first. From the Chapman–Kolmogorov equation again

$$p_{ii}^{(m+r)} \geqslant p_{ij}^{(m)} p_{ji}^{(r)} = ab > 0.$$

Hence if d_i is the period of state i, $m+r$ must be a multiple of d_i. Then the left-hand side of (4.5.10) vanishes unless n is also a multiple of d_i implying in turn that $p_{jj}^{(n)} = 0$ unless n is a multiple of d_i. We have therefore that if d_j is the period of j then for some positive integer k_1

$$d_j = k_1 d_i.$$

Interchanging i and j the same argument yields

$$d_i = k_2 d_j$$

for a positive integer k_2. But these two equalities cannot be simultaneously satisfied for positive integers k_1, k_2 unless $k_1 = k_2 = 1$. Hence $d_i = d_j$ It remains to show that any two given states i and j are both either transient, null recurrent, or positive recurrent. Equations (4.5.10), (4.5.11) imply that both sums $\sum_n p_{ii}^{(n)}$ and $\sum_n p_{jj}^{(n)}$ are convergent or divergent together. From Definitions 4.5.1. and 4.5.2 it follows that i and j are either both transient or both recurrent.

Suppose finally that i is null recurrent. Then since m and r are fixed $\lim_{n\to\infty} p_{ii}^{(n+m+r)} = 0$ implies $\lim_{n\to\infty} p_{jj}^{(n)} = 0$ from (4.5.10). Conversely if j is null recurrent then by (4.5.11) i must also be null recurrent. This completes the proof. □

Theorem 4.5.4 is an important example of a 'solidarity' result. In examining an irreducible chain for periodicity and recurrence it suffices therefore to consider only one particular state. In future we will say that an irreducible chain is periodic, transient, null recurrent, or positive recurrent according to the classification of any given state.

The next proposition highlights an important distinction between transient and recurrent irreducible chains. We first require a preliminary result

Lemma. For all pairs (i, j) of states

$$F_{ij} = p_{ij} - p_{ij}F_{jj} + \sum_k p_{ik}F_{kj} \tag{4.5.12}$$

Proof. By definition, for $n = 2, 3, \ldots,$

$$\begin{aligned}
f_{ij}^{(n)} &= \Pr(X_n = j; X_\nu \neq j, \nu = 1, 2, \ldots, n-1 \,|\, X_0 = i) \\
&= \Pr\Bigg(\bigcup_{\substack{k \\ k \neq j}} \{X_1 = k; X_\nu \neq j, \nu = 2, 3, \ldots, n-1; X_n = j\} \,|\, X_0 = i\Bigg) \\
&= \sum_{\substack{k \\ k \neq j}} \Pr(X_1 = k; X_\nu \neq j, \nu = 2, 3, \ldots, n-1; X_n = j \,|\, X_0 = i) \\
&= \sum_{\substack{k \\ k \neq j}} \Pr(X_1 = k \,|\, X_0 = i)\Pr(X_\nu \neq j, \nu = 2, 3, \ldots, n-1; \\
&\qquad\qquad X_n = j \,|\, X_0 = i, X_1 = k) \\
&= \sum_{\substack{k \\ k \neq j}} p_{ik} f_{kj}^{(n-1)}
\end{aligned}$$

since

$$\begin{aligned}
\Pr(X_\nu \neq j, \nu = 2, 3, \ldots, n-1; X_n = j \,|\, X_0 = i, X_1 = k) \\
= \Pr(X_\nu \neq j, \nu = 1, 2, \ldots, n-2; X_{n-1} = j \,|\, X_0 = k) \\
= f_{kj}^{(n-1)}.
\end{aligned}$$

□

Theorem 4.5.5 If an irreducible chain is recurrent then

$$F_{ij} = \sum_{n=1}^{\infty} f_{ij}^{(n)} = 1$$

for all states i and j. On the other hand if the chain is transient there exists at least one pair (k, j) for which $F_{kj} < 1$.

Proof. Suppose the chain is recurrent. Then $F_{jj} = 1$ and (4.5.12) reduces to*

$$F_{ij} = \sum_k p_{ik} F_{kj}. \tag{4.5.13}$$

Multiply both sides by p_{li} and sum over i to obtain

$$\begin{aligned} F_{lj} &= \sum_i p_{li} F_{ij} \\ &= \sum_k \left(\sum_i p_{li} p_{ik}\right) F_{kj} \\ &= \sum_k p_{lk}^{(2)} F_{kj}. \end{aligned}$$

Repeating this procedure m times we find the useful extension

$$F_{ij} = \sum_k p_{ik}^{(m)} F_{kj} \tag{4.5.14}$$

for recurrent chains, $m = 1, 2, \ldots$. Then taking $i = j$ in (4.5.14) we find

$$0 = 1 - F_{ii} = \sum_k p_{ik}^{(m)} (1 - F_{ki})$$

since $\sum_k p_{ik}^{(m)} = 1$ for every fixed m. By positivity this implies $F_{ki} = 1$ for every k for which $p_{ik}^{(m)} > 0$. But for an irreducible chain there exists an m such that $p_{ik}^{(m)} > 0$ for every pair (i, k). Hence for irreducible recurrent chains $F_{ki} = 1$ for all k and i. To prove the second part of the theorem suppose the chain is transient. The general identity (4.5.12) continues to hold and in Exercise 9 you are asked to deduce

$$(1 - F_{ii}) \sum_{n=0}^{m} p_{ii}^{(n)} = \sum_k p_{ik}^{(m)} (1 - F_{ki}), \qquad m = 1, 2, \ldots. \tag{4.5.15}$$

* Since $\sum_k p_{ik} = 1$ a solution to (4.5.13) is $F_{kj} \equiv 1$, but we cannot assume uniqueness.

By transience the left-hand side is a finite non-zero quantity for every $m > 0$. Hence the right-hand side cannot vanish and the conclusion that at least one $F_{ki} < 1$ follows. This completes the proof of the theorem. □

Corollary In an aperiodic irreducible positive recurrent chain

$$\lim_{n\to\infty} p_{ij}^{(n)} = 1/\mu_{jj}$$

for all states j, independently of the initial state i.

Proof. From Theorem 4.5.3 (ii)

$$\begin{aligned}\lim_{n\to\infty} p_{ij}^{(n)} &= F_{ij}/\mu_{jj}\\ &= 1/\mu_{jj} \quad \text{by Theorem 4.5.5.}\end{aligned}$$ □

The important point of the theorem is that it is only for recurrent irreducible chains that the probability is one the every state will be visited irrespective of the initial state.

The next result concerns chains with a finite state space which, without loss of generality, we will suppose to be the K states $0, 1, \ldots, K-1$.

Theorem 4.5.6 All states of an aperiodic finite irreducible chain are positive recurrent. Supposing the state space to be $\{0, 1, \ldots, K-1\}$ the limits

$$\pi_j = \lim_{n\to\infty} p_{ij}^{(n)} = 1/\mu_{jj}, \qquad j = 0, 1, \ldots, K-1$$

constitute a probability distribution (i.e. $\sum_{j=0}^{K-1} \pi_j = 1$) and are the unique (up to a multiplicative constant) solutions of the equations

$$\pi_j = \sum_{k=0}^{K-1} \pi_k p_{kj}, \qquad j = 0, 1, \ldots, K-1. \tag{4.5.16}$$

Proof. Since $\sum_{j=0}^{K-1} p_{ij}^{(n)} = 1$ for every finite n we have

$$1 = \lim_{n\to\infty} \sum_{j=0}^{K-1} p_{ij}^{(n)} = \sum_{j=0}^{K-1} \lim_{n\to\infty} p_{ij}^{(n)} = \sum_{j=0}^{K-1} \pi_j, \tag{4.5.17}$$

the interchange of limit with summation being permissible since K is finite. If the chain is transient or null recurrent, $p_{ij}^{(n)} \to 0$ for all i and j which contradicts (4.5.16). Hence (4.5.17) holds if and only if the chain is positive recurrent. We infer that the chain is positive recurrent with $\Sigma \pi_i = 1$. Note that taking limits in the Chapman–Kolmogorov equation

$$p_{ij}^{(n+1)} = \sum_{k=0}^{K-1} p_{ik}^{(n)} p_{kj}$$

yields

$$\pi_j = \sum_{k=0}^{K-1} \pi_k p_{kj}, \qquad j = 0, 1, \ldots, K-1,$$

so that the limits π_j certainly satisfy (4.5.16). To prove uniqueness suppose that $\{u_j\}$ is any other solution of (4.5.16). Then multiplying through by p_{jl} and summing over j

$$u_l = \sum_j u_j p_{jl} = \sum_j \sum_k u_k p_{kj} p_{jl} = \sum_k u_k p_{kl}^{(2)}.$$

Repeating this procedure m times yields

$$u_j = \sum_{k=0}^{K-1} u_k p_{kj}^{(m)}.$$

Taking the limit as $m \to \infty$ we have, by positive recurrence,

$$u_j = \sum_{k=0}^{K-1} u_k \pi_j = \pi_j \sum_{k=0}^{K-1} u_k.$$

which establishes the required uniqueness. □

A restatement of Theorem 4.5.6 is that every aperiodic irreducible chain on a finite state space is convergent in distribution, the limit distribution being given by the $\{\pi_j\}$ which satisfy (4.5.16). The situation is more complicated when the state space is infinite and these chains are by no means necessarily positive recurrent. Indeed trying to extend the proof of Theorem 4.5.6 to the case $K = \infty$ one notes that the key equation (4.5.17) is generally false since

$$1 = \lim_{n \to \infty} \sum_{j=0}^{\infty} p_{ij}^{(n)} \geqslant \sum_{j=0}^{\infty} \lim_{n \to \infty} p_{ij}^{(n)}.$$

We only state the general result.

Theorem 4.5.7 Suppose the chain is irreducible and aperiodic with state space $\{0, 1, 2, \ldots\}$. The necessary and sufficient condition for $\pi_j = \lim_{n\to\infty} p_{ij}^{(n)}, j = 0, 1, \ldots,$ to constitute a probability distribution is that the chain be positive recurrent. If this is the case the π_j are the unique (up to a multiplicative constant) solutions of the equations

$$\pi_j = \sum_{i=0}^{\infty} \pi_i p_{ij}, \qquad j = 0, 1, \ldots, \tag{4.5.18}$$

and are related to the recurrence time distributions by

$$\pi_j = 1/\mu_{jj}. \qquad \square$$

To sum up our discussion: the primary distinction is between transient and recurrent chains (c.f. Theorem 4.5.5), and then between those that are null recurrent or positive recurrent. It is only in the last case that one has convergence in distribution. The distinction between transience and recurrence is explored further in § 5.5.

Example 1 Simple random walk on the non-negative integers with a reflecting barrier at the origin (c.f. Example 3.5.4 and Exercises 3.5.4–7). The transition matrix is

$$\mathscr{P} = \begin{pmatrix} q & p & 0 & 0 & 0 & 0 & \ldots \\ q & 0 & p & 0 & 0 & 0 & \ldots \\ 0 & q & 0 & p & 0 & 0 & \ldots \\ 0 & 0 & q & 0 & p & 0 & \ldots \\ \ldots & \ldots & \ldots & \ldots & \ldots & \ldots & \ldots \end{pmatrix}$$

The chain is clearly irreducible and aperiodic. Its recurrence classification is therefore the same as that on any one state, say 0. To find F_{00} note that the first passage time

$$U_1(z) = \{1-(1-4pqz^2)^{\frac{1}{2}}\}/2pz$$

of Exercise 3.5.7 is precisely $U_1(z) = F_{10}(z)$. But

$$f_{00}^{(1)} = p_{00} = q$$
$$f_{00}^{(n)} = p_{01} f_{10}^{(n-1)} = pf_{10}^{(n-1)}, \qquad n = 2, 3, \ldots$$

Hence

$$F_{00}(z) = qz + pz\,F_{10}(z) = qz + \tfrac{1}{2}\{1-(1-4pqz^2)^{\frac{1}{2}}\}.$$

Since $(1-4pq)^{\frac{1}{2}} = (1-4p+4p^2)^{\frac{1}{2}} = |1-2p|$,

$$F_{00} = \lim_{z \to 1-} F_{00}(z) = q+\tfrac{1}{2}(1-|1-2p|).$$

Thus

$$F_{00} = \begin{cases} q+\tfrac{1}{2}\{1-(2p-1)\} = 2q < 1 & \text{if } p > q \\ q+\tfrac{1}{2}\{1-(1-2p)\} = q+p = 1 & \text{if } p \leqslant q. \end{cases}$$

The chain is therefore recurrent if $p \leqslant q$ and transient if $p > q$. When $p \leqslant q$ the mean recurrence time μ_{00} is given by

$$\begin{aligned} \mu_{00} &= \lim_{z \to 1-} F'_{00}(z) \\ &= \lim_{z \to 1-} \left[q + \frac{2pqz}{(1-4pqz^2)^{\frac{1}{2}}} \right] \\ &= q + \frac{2pq}{|1-2p|} \\ &= \begin{cases} \infty & \text{if } p = \tfrac{1}{2} = q, \\ q/(q-p) & \text{if } p < q, \end{cases} \end{aligned}$$

and we have positive recurrence if $p < q$, null recurrence if $p = q$. This Example is continued in Exercise 4. □

Exercises

1. *Unrestricted simple random walk.* $p_{ii+1} = p$, $p_{ii-1} = 1-p = q$, $p_{ij} = 0$ otherwise, $i, j = 0, \pm 1, \pm 2, \ldots$. Show that

$$\begin{aligned} p_{ij}^{(n)} = p_{0j-i}^{(n)} &= \binom{n}{\tfrac{1}{2}(n+j-i)} p^{\frac{1}{2}(n+j-i)} q^{\frac{1}{2}(n-j+i)} \quad \text{if } n+j-i \text{ even} \\ &= 0 \quad \text{if } n+j-i \text{ odd}. \end{aligned}$$

Conclude that this chain is irreducible with period 2. Use Stirling's approximation (2.5.15) to show that

$$p_{ii}^{(2n)} \sim \frac{(4pq)^n}{(2\pi n)^{\frac{1}{2}}} \quad \text{as } n \to \infty,$$

and hence that the chain is recurrent if $p = q = \frac{1}{2}$, transient if $p \neq q$.

2. (Continuation.) Show that the generating functions of the first passage distributions are

$$F_{ii}(z) = F_{00}(z) = 1 - (1 - 4pqz^2)^{\frac{1}{2}},$$

$$F_{ii+1}(z) = F_{01}(z) = \frac{1}{2qz}\{1-(1-4pqz^2)^{\frac{1}{2}}\},$$

$$F_{ii-1}(z) = F_{10}(z) = \frac{1}{2pz}\{1-(1-4pqz^2)^{\frac{1}{2}}\},$$

$$F_{ij}(z) = [F_{01}(z)]^{j-i} \quad \text{if } j > i, \quad F_{ij}(z) = [F_{10}(z)]^{i-j} \quad \text{if } j < i.$$

Verify the conditions for recurrence found in Exercise 1 and show that this chain can never be positive recurrent.

3. (Continuation.) The walk is said to have *positive drift* if $p > q$, *negative drift* if $p < q$, and no drift if $p = q$. For walks with positive drift show that for all i,

$$F_{ij} = 1, \qquad j = i+1, i+2, \ldots,$$
$$F_{ij} < 1, \qquad j = i-1, i-2, \ldots,$$

and state the corresponding result when drift is negative. Note that for walks with no drift (the recurrent case) $F_{ij} = 1$, all i, j.

4. (Continuation of Example 1.) Find $F_{01}(z), F_{ii+1}(z)$ and using the fact that the chain can increase by only unit amounts show that

$$F_{0j}(z) = \prod_{k=0}^{j-1} F_{kk+1}(z).$$

5. (Continuation.) Solve (4.5.18) for this walk to find the limits $\pi_j = \lim_{n \to \infty} p_{ij}^{(n)}$ when the chain is positive recurrent.

6. Suppose the transition matrix of Example 1 is truncated by imposing a reflecting barrier at state $K-1$. Solve (4.5.16) to find $\{\pi_j, j = 0, 1, \ldots, K-1\}$ and verify that it is a probability distribution irrespective of the relative magnitudes of p and q.

7. Show that if an irreducible chain has $p_{ii} > 0$ for at least one i then the chain must be aperiodic.

8. Is the chain defined by the matrix

$$\mathscr{P} = \begin{pmatrix} 0 & 0 & p & q & 0 \\ 0 & 0 & q & p & 0 \\ 0 & 0 & 0 & 0 & 1 \\ p & q & 0 & 0 & 0 \\ q & p & 0 & 0 & 0 \end{pmatrix}$$

periodic? If so find its period.

9. Establish (4.5.15).

10. The gambler's ruin problem of Exercises 3.5.5–7 (restricted simple random walk with absorbing barriers at the end points) is a case of a reducible Markov chain since starting from 0 or b no communication is possible with the other states. Show that in the modification of (3.5.12)

$$\mathscr{P} = \begin{pmatrix} q & p & 0 & 0 & 0 & . & . & . & . \\ q & p & 0 & 0 & 0 & . & . & . & . \\ 0 & q & 0 & p & 0 & . & . & . & . \\ 0 & 0 & q & 0 & p & . & . & . & . \\ . & . & . & . & . & 0 & q & 0 & p \\ . & . & . & . & . & 0 & q & 0 & p \\ . & . & . & . & . & . & 0 & 0 & 1 \end{pmatrix}$$

the state space reduces to the three classes $\{0, 1\}$, $\{b\}$, $\{2, 3, \ldots, b-1\}$, the first two being recurrent and the third transient. Find $\lim_{n\to\infty} p_{00}^{(n)}$.

11. A set of states from which no escape is possible is called *closed* (e.g. $\{0, 1\}$ in the preceding exercise). Suppose the state space contains the closed set $\{0, 1, \ldots, a-1\}$. Verify that the transition matrix can be written in partitioned form

$$\mathscr{P} = \begin{pmatrix} A & 0 \\ B & C \end{pmatrix}$$

where A is an $a \times a$ stochastic matrix, and that the n-step transition probabilities are the elements of

$$\mathscr{P}^n = \begin{pmatrix} A^n & 0 \\ B_n & C^n \end{pmatrix}$$

Hence if $i, j \in \{0, 1, \ldots, a-1\}$ then $p_{ij}^{(n+m)} = \sum_{k=0}^{a-1} p_{ik}^{(n)} p_{kj}^{(m)}$.

12. Let $p^{(0)} = 1, p^{(n)} = \sum_{\nu=1}^{n} f^{(\nu)} p^{(n-\nu)}, n = 1, 2, \ldots,$ and suppose $\{f^{(n)}\}$ is the geometric distribution $f^{(n)} = (1-\delta)\delta^{n-1}, n = 1, 2, \ldots$. Show that for $n \geqslant 1, p^{(n)}$ is independent of n and in fact $p^{(n)} = 1-\delta$, $n = 1, 2, \ldots$ Prove the converse proposition that if $p^{(0)} = 1, p^{(n)} = c$, $n = 1, 2, \ldots$, then

$$c = f^{(n)} \Big/ \sum_{\nu=n}^{\infty} f^{(\nu)}$$

and $\{f^{(n)}\}$ must be the geometric distribution $f^{(n)} = c(1-c)^{n-1}$, $n = 1, 2, \ldots$

13. (Continuation.) More generally let $f^{(n)} = \Pr(Y = n)$. Then the ratio

$$c_n = f^{(n)} \Big/ \sum_{\nu=n}^{\infty} f^{(\nu)} = \Pr(Y = n \mid Y \geqslant n)$$

is the discrete version of the force of mortality of Exercise 3.2.9. Interpreting Y as a lifetime, conclude from Exercise 12 that geometrically distributed lifetimes are the only discrete ones with constant force of mortality.

14. (Continuation.) In the stochastic matrix on $\{1, 2, \ldots\}$

$$\mathscr{P} = \begin{pmatrix} q_1 & p_1 & 0 & 0 & 0 & . & . & \\ q_2 & 0 & p_2 & 0 & 0 & . & . & . \\ q_3 & 0 & 0 & p_3 & 0 & . & . & . \\ q_4 & 0 & 0 & 0 & p_4 & . & . & . \\ . & . & . & . & . & . & . & . \end{pmatrix}$$

with $p_j + q_j = 1$, let $p_j = \sum_{v=j+1}^{\infty} f^{(v)} \Big/ \sum_{v=j}^{\infty} f^{(v)}$. Show that

$$p_{11}^{(n)} = \sum_{v=1}^{n} f^{(v)} p_{11}^{(n-v)}, \qquad n = 1, 2, \ldots,$$

implying that the results the two preceding exercises apply to $p_{11}^{(n)}$. Find $\lim_{n\to\infty} p_{1j}^{(n)}$.

15. The name stationary or invariant distribution for the solution $\{\pi_j\}$ of (4.5.16) or (4.5.18) is appropriate since if the initial distribution is $\Pr(X_0 = i) = \pi_i$, $i = 0, 1, \ldots$, then for the absolute probabilities

$$\Pr(X_n = j) = \sum_{i=0}^{\infty} \Pr(X_0 = i) p_{ij}^{(n)} = \pi_j$$

independently of n. Verify this statement.

Chapter 5
ALMOST SURE CONVERGENCE

The preceding chapter concerned convergence in the usual sense of sequences of probabilities or sequences of distribution functions. Convergence properties of a sequence of random variables $\{X_n)$ were described in terms of the limit behaviour of the associated quantities $\Pr(|X_n - X| < \varepsilon)$ or $\Pr(X_n \leqslant x)$. Now a random variable is a function from a sample space $\Omega = \{\omega\}$ into the real line. It is therefore natural to consider $\{X_n\}$ itself as a sequence of such functions and limit problems that arise in this context fall under the general heading of what is commonly called *strong convergence.*

Attention is focused on *almost sure* (a.s.) convergence, formally introduced in Definition 5.2.1, in which the central issue is the probability assigned to sets of the form

$$\{\omega \mid \lim_{n\to\infty} X_n(\omega) = X(\omega)\}.$$

Such questions make sense only of all members of the sequence $\{X_n\}$ and the limit X (if there is one) are random variables (functions) on the same probability space $(\Omega, \mathscr{F}, P)$. It sometimes happens that whilst it is clear on which space any finite number of the X_n are defined, this is not necessarily so for the whole sequence.

A case in point arises when the X_j are defined (or induced) by specification of the joint distribution functions $\Pr(X_1 \leqslant x_1, X_2 \leqslant x_2, \ldots, X_n \leqslant x_n)$ for every finite $n = 1, 2, \ldots$, and on the basis of this information it is required to find $\Pr(\lim_{n\to\infty} X_n = X)$, say. If $(\Omega_j, \mathscr{F}_j, P_j)$ is the space induced by $\Pr(X_j \leqslant x_j)$ then, as in § 3.1, $X_1, X_2, \ldots, X_n$ are jointly defined on

$$\begin{aligned}\Omega^{(n)} &= \Omega_1 \times \Omega_2 \times \ldots \times \Omega_n \\ &= \{\omega^{(n)} = (w_1, w_2, \ldots, w_n) | \, w_j \in \Omega_j, j = 1, 2, \ldots, n\}\end{aligned}$$

The probability function $P^{(n)}$ on $\Omega^{(n)}$ is prescribed by

$$P^{(n)}(\{\omega^{(n)} \mid X_1 \leqslant x_1, x_2 \leqslant x_2, \ldots, X_n \leqslant x_n\}) \\ = \Pr(X_1 \leqslant x_1, X_2 \leqslant x_2, \ldots, X_n \leqslant x_n),$$

and is consistent in the sense that all marginal distributions can be obtained from it. Furthermore probabilities remain the same under permutation of the arguments; thus for example

$$P^{(n)}(\{\omega^{(n)} \mid X_2 \leqslant x_2, X_1 \leqslant x_1, \ldots, X_n \leqslant x_n\}) \\ = P^{(n)}(\{\omega^{(n)} \mid X_1 \leqslant x_1, X_2 \leqslant x_2, \ldots, X_n \leqslant x_n\}).$$

As $n \to \infty$, $\Omega^{(n)} \to \Omega$ where

$$\Omega = \{\omega = (w_1, w_2, \ldots) \mid w_j \in \Omega_j, j = 1, 2, \ldots\}$$

is the sample space of infinite vectors on which all members of $\{X_n\}$ are defined. The question arises: is there a function P assigning probabilities to events on Ω such that P is consistent with the $P^{(n)}$, $n = 1, 2, \ldots$, defined by the joint distribution functions of only finitely many variables? If there is such a P then all the X_n and limit, if any, are defined on the same probability space $(\Omega, \mathscr{F}, P)$.

The answer to the question is in the affirmative. Indeed there is a unique such P, and this result is known as the *Kolmogorov Extension Theorem* the truth of which we assume although the proof is omitted. Thus whenever necessary (§§ 3–5) we assume the existence of a probability space $(\Omega, \mathscr{F}, P)$ on which all members of $\{X_n\}$ are defined.

The notion of almost sure convergence is intimately related to a certain repetitive behaviour on the part of members of $\{X_n\}$. The intent of § 1 is to clarify what is meant by the probability of events occurring infinitely often and the basic results concerning random variables are in § 2. §§ 3, 4 treat the classical strong law of large numbers, a result of fundamental importance. The final section discusses some aspects of Markov chains in the setting of almost sure convergence.

5.1 INFINITE SEQUENCES OF EVENTS

Suppose $\{A_n, n = 1, 2, \ldots\}$ is a sequence of events defined on a probability space $(\Omega, \mathscr{F}, P)$. The limit of a monotone sequence $\{A_n\}$

was defined in the Appendix to Chapter 1 to be

$$\lim_{n\to\infty} A_n = \begin{cases} \bigcap_{n=1}^{\infty} A_n & \text{if } A_1 \supseteq A_2 \supseteq A_3 \supseteq \ldots, \\ \bigcup_{n=1}^{\infty} A_n & \text{if } A_1 \subseteq A_2 \subseteq A_3 \subseteq \ldots. \end{cases} \tag{5.1.1}$$

The theorem of that Appendix established for such monotone sequences the result

$$\lim_{n\to\infty} P(A_n) = P(\lim_{n\to\infty} A_n). \tag{5.1.2}$$

Our present concern is to examine the limit properties of sequences $\{B_n\}$ that are not necessarily monotone. In the next section we extend the discussion to sequences of random variables.

For any sequence $\{B_n\}$ of events in $\mathscr{F}$ the unions

$$A_n = \bigcup_{j=n}^{\infty} B_j, \qquad n = 1, 2, 3, \ldots,$$

constitute a monotone sequence $\{A_n\}$

$$\text{(since } A_n = \bigcup_{j=n}^{\infty} B_j \supseteq \bigcup_{j=n+1}^{\infty} B_j = A_{n+1}\text{).}$$

Hence by (5.1.1) the limit

$$\lim_{n\to\infty} A_n = \bigcap_{n=1}^{\infty} \bigcup_{j=n}^{\infty} B_j$$

is always defined and is called the *limit superior* (or upper limit) of the sequence $\{B_n\}$; written

$$\limsup_{n\to\infty} B_n = \bigcap_{n=1}^{\infty} \bigcup_{j=n}^{\infty} B_j = \mathrm{B}. \tag{5.1.3}$$

B is the set of sample points ω that for each $n = 1, 2, \ldots$ belong to some $B_j, j \geqslant n$; that is B contains those ω that belong to infinitely many of $B_1, B_2, \ldots$. For this reason B can be called the event 'members of the sequence $\{B_n\}$ occur infintely often' or 'infinitely many of $B_1, B_2, \ldots$ occur'. Abbreviating infinitely often to i.o. a common notation is

$$\limsup_{n\to\infty} B_n = \{\omega \mid \omega \in B_n \text{ i.o.}\} = \{\omega \mid \omega \in \bigcap_{n=1}^{\infty} \bigcup_{j=n}^{\infty} B_j\}.$$

In the same way the intersections

$$C_n = \bigcap_{j=n}^{\infty} B_j, \qquad n = 1, 2, \ldots,$$

form a non-decreasing sequence, $C_1 \subseteq C_2 \subseteq C_3 \subseteq \ldots$, and the limit

$$\lim_{n\to\infty} C_n = \bigcup_{n=1}^{\infty}\bigcap_{j=n}^{\infty} B_j$$

is called the *limit inferior* (or lower limit) of $\{B_n\}$.

$$\liminf_{n\to\infty} B_n = \bigcup_{n=1}^{\infty}\bigcap_{j=n}^{\infty} B_j = \underline{B}. \tag{5.1.4}$$

$\underline{B}$ is the set of ω which for some $n \geqslant 1$ belong to all $B_j, j \geqslant n$. Put another way, $\underline{B}$ is the set of sample points that belong to all except possibly a finite number of the events B_j.

If $B = \underline{B}$ then we say that $\{B_n\}$ is *convergent* and $\lim_{n\to\infty} B_n$ is given by the common value

$$B = \lim_{n\to\infty} B_n = \limsup_{n\to\infty} B_n = \liminf_{n\to\infty} B_n. \tag{5.1.5}$$

Example 1 Ω is the real line, $B_n = \{\omega \mid a+\alpha_n < \omega < b+\beta_n\}$ where both α_n and $\beta_n \to 0$ as $n \to \infty$. Then if the α_n and β_n are non-negative

$$\lim_{n\to\infty} B_n = \{\omega \mid a < \omega \leqslant b\},$$

which is the set containing those ω belonging to infinitely many B_j. □

Example 2

$$B_n = \begin{cases} \{\omega \mid 1-n^{-1} \leqslant \omega \leqslant 3+n^{-1}\}, & n \text{ odd,} \\ \{\omega \mid 2-n^{-1} \leqslant \omega \leqslant 5+n^{-1}\}, & n \text{ even} \end{cases}$$

Then

$$\limsup_{n\to\infty} B_n = \bigcap_{n=1}^{\infty}\bigcup_{j=n}^{\infty} B_j = \{\omega \mid 1 \leqslant \omega \leqslant 5\}$$

and

$$\liminf_{n\to\infty} B_n = \bigcup_{n=1}^{\infty}\bigcap_{j=n}^{\infty} B_j = \{\omega \mid 2 \leqslant \omega \leqslant 3\}.$$

$\lim_{n\to\infty} B_n$ does not exist since the upper and lower limits are not equal. □

Example 3 Suppose $\{B_n\}$ is a sequence of disjoint events. Then $\liminf_{n\to\infty} B_n = \emptyset$ since $\bigcap_{j=n}^{\infty} B_j = \emptyset$ for every n. Turning to $\limsup_{n\to\infty} B_n$, recall that it is the event containing these ω which belong to infinitely many B_j. But, since the B_j are disjoint, $\omega \in B_k$ implies $\omega \notin B_j$ for all $j \neq k$, i.e. a given sample point belongs to one and only one B_j, if it belongs to any. Hence $\limsup_{n\to\infty} B_n = \emptyset$ and for disjoint events we have $\lim_{n\to\infty} B_n = \emptyset$. □

Note that since $\mathscr{F}$ is a σ field (and therefore closed under countable unions and intersections) both the limit inferior and the limit superior of $\{B_n\}$ are members of $\mathscr{F}$ and can thus be assigned probabilities by the function P. Also

$$\liminf_{n\to\infty} B_n \subseteq \limsup_{n\to\infty} B_n,$$

since if $\omega \in \liminf_{n\to\infty} B_n$ it is an element of all but a finite number of the B_j and hence belongs to infinitely many of them. Consequently

$$P(\liminf_{n\to\infty} B_n) \leqslant P(\limsup_{n\to\infty} B_n).$$

However it is possible to extend this inequality as follows;

Theorem 5.1.1 If $\{B_n\}$ is a sequence of events on $(\Omega, \mathscr{F}, P)$ then

$$P(\liminf_{n\to\infty} B_n) \leqslant \liminf_{n\to\infty} P(B_n) \leqslant \limsup_{n\to\infty} P(B_n) \leqslant P(\limsup_{n\to\infty} B_n). \quad (5.1.6)$$

If B_n is convergent then

$$\lim_{n\to\infty} P(B_n) = P(\lim_{n\to\infty} B_n). \quad (5.1.7)$$

Proof: Consider first the left hand inequality of (5.1.6). From (5.1.2)

$$\lim_{n\to\infty} P(\bigcap_{j=n}^{\infty} B_j) = P(\bigcup_{n=1}^{\infty}\bigcap_{j=n}^{\infty} B_j) = P(\liminf_{n\to\infty} B_n).$$

But for each n, $\bigcap_{j=n}^{\infty} B_j \subseteq B_n$. Hence

$$P(\bigcap_{j=n}^{\infty} B_j) \leqslant P(B_n)$$

and

$$P(\liminf_{n\to\infty} B_n) = \lim_{n\to\infty} P(\bigcap_{j=n}^{\infty} B_j) \leqslant \liminf_{n\to\infty} P(B_n).$$

Turning to the right hand inequality in (5.1.6) we have again from (5.1.2)

$$\lim_{n\to\infty} P(\bigcup_{j=n}^{\infty} B_j) = P(\bigcap_{n=1}^{\infty}\bigcup_{j=n}^{\infty} B_j) = P(\limsup_{n\to\infty} B_n).$$

But

$$P(B_n) \leqslant P(\bigcup_{j=n}^{\infty} B_j)$$

since $B_n \subseteq \bigcup_{j=n}^{\infty} B_j$, and hence

$$\limsup_{n\to\infty} P(B_n) \leqslant \lim_{n\to\infty} P(\bigcup_{j=n}^{\infty} B_j) = P(\limsup_{n\to\infty} B_n).$$

Finally, since $P(B_n)$ is an ordinary sequence of real numbers we always have

$$\liminf_{n\to\infty} P(B_n) \leqslant \limsup_{n\to\infty} P(B_n),$$

which completes the proof of (5.1.6).

For a convergent sequence (5.1.7) is an immediate consequence of (5.1.6), for then

$$\limsup_{n\to\infty} B_n = \liminf_{n\to\infty} B_n = \lim_{n\to\infty} B_n,$$

implying that the extreme members of the inequality (5.1.6) are equal. In that case (5.1.6) reduces to equality throughout and $\lim_{n\to\infty} P(B_n)$ is by definition the common value of $\liminf_{n\to\infty} P(B_n)$ and $\limsup_{n\to\infty} P(B_n)$. □

One possible use of the theorem is the evaluation of the probability that infinitely many of $B_1, B_2, \ldots,$ occur, i.e. the evaluation of $P(\limsup_{n\to\infty} B_n)$, or $P(\lim_{n\to\infty} B_n)$ in the convergent case, from the individual probabilities $P(B_n)$.

The next theorem (the two parts of which are commonly known as the first and second Borel–Cantelli Lemmas respectively) is particularly useful.

Theorem 5.1.2 (Borel–Cantelli). Suppose $\{B_n\}$ is a sequence of events on $(\Omega, \mathscr{F}, P)$. Then

(i) $\sum_{n=1}^{\infty} P(B_n) < \infty$ implies $P(\limsup_{n\to\infty} B_n) = 0$,

(ii) if the $\{B_n\}$ are mutually independent and $\sum_{n=1}^{\infty} P(B_n) = \infty$ then $P(\limsup_{n\to\infty} B_n) = 1$.

Proof. (i) $\sum_{j=1}^{\infty} P(B_j) < \infty$ implies $\lim_{n\to\infty} \sum_{j=n}^{\infty} P(B_j) = 0$. But for every n

$$\limsup_{n\to\infty} B_n = \bigcap_{n=1}^{\infty}\bigcup_{j=n}^{\infty} B_j \subseteq \bigcup_{j=n}^{\infty} B_j. \tag{5.1.8}$$

Therefore

$$\begin{aligned} P(\limsup_{n\to\infty} B_n) &\leqslant P(\bigcup_{j=n}^{\infty} B_j) \\ &\leqslant \sum_{j=n}^{\infty} P(B_j) \to 0 \qquad \text{as } n \to \infty. \end{aligned}$$

(ii) We show that the hypotheses of the theorem imply

$$P(\bigcup_{j=n}^{\infty} B_j) = 1$$

for each n. For given n and N (see Exercise 1.2.3(f))

$$(\bigcup_{j=n}^{N} B_j) \cup (\bigcap_{j=n}^{N} \bar{B}_j) = \Omega.$$

Then $P(\bigcup_{j=n}^{\infty} B_j) \geqslant P(\bigcup_{j=n}^{N} B_j)$ implies

$$\begin{aligned} 1 - P(\bigcup_{j=n}^{\infty} B_j) &\leqslant 1 - P(\bigcup_{j=n}^{N} B_j) \\ &= P(\bigcap_{j=n}^{N} \bar{B}_j) \\ &= \prod_{j=n}^{N} P(\bar{B}_j), \qquad \text{by independence,} \\ &= \prod_{j=n}^{N} (1 - P(B_j)) \\ &\leqslant \exp\{-\sum_{j=n}^{N} P(B_j)\}, \end{aligned}$$

since $1-x \leqslant e^{-x}$ for $0 < x < 1$. Let $N \to \infty$. Then for each fixed n

$$1-P(\bigcup_{j=n}^{\infty} B_j) \leqslant \lim_{N\to\infty} \exp\{-\sum_{j=n}^{N} P(B_j)\}$$

$$= 0 \qquad \text{since } \sum_{j=n}^{\infty} P(B_j) = \infty.$$

Hence $P(\bigcup_{j=n}^{\infty} B_j) = 1$ for each n. But this implies $\lim_{n\to\infty} P(\bigcup_{j=n}^{\infty} B_j) = 1$

and hence $P(\limsup_{n\to\infty} B_n) = P(\bigcap_{n=1}^{\infty}\bigcup_{j=n}^{\infty} B_j) = \lim_{n\to\infty} P(\bigcup_{j=n}^{\infty} B_j) = 1.$ □

Exercises

1. A sequences $\{b_n, 1, 2, \ldots\}$ is convergent to b if for every $\varepsilon > 0$ there exists $N = N(\varepsilon)$ such that $|b_n - b| < \varepsilon$ for all $n > N$. Show that this is equivalent to requiring that both

$$\limsup_{n\to\infty} b_n = \inf_{n\geqslant 1} \sup_{j\geqslant n} b_j$$

and

$$\liminf_{n\to\infty} b_n = \sup_{n\geqslant 1} \inf_{j\geqslant n} b_j$$

be equal to b.

2. Consider the interval sets B_n of Example 2. If $\Omega = \bigcup_{n=1}^{\infty} B_n$ and $P(B_n) = (\text{length of } B_n)/(\text{length of } \Omega)$ find $P(\liminf_{n\to\infty} B_n)$, $P(\limsup_{n\to\infty} B_n)$. $\limsup_{n\to\infty} P(B_n)$, $\liminf_{n\to\infty} P(B_n)$, and show that in this case the inequalities in (5.1.6) are strict.

3. For k an integer $\geqslant 2$, let B_n be the event that a run of n successes occurs between the k^nth and k^{n+1}th members of a sequences of Bernoulli trials. If p is the probability of success on any one trial, for what values of p is the probability 1 that infinitely many B_n occur?

4. Let $\Omega = \{\omega | 0 \leqslant \omega \leqslant 1\}$ with the probabilities of interval sets equal to their length. If $A_n = \{\omega | 0 \leqslant \omega \leqslant 1 - 1/3^n\}$ and $B_n = \bar{A}_n A_{2n}$, $n = 1, 2, \ldots$, show that $P\limsup_{n\to\infty} B_n) = 0$.

5. $Y_1, Y_2, \ldots,$ are independent random variables indentically distributed as Y whose distribution is $\Pr(Y = -1) = 1-p, \Pr(Y = 1) = p$. If $S_n = Y_1 + Y_2 + \ldots + Y_n$, show that provided $p \neq \frac{1}{2}$ the probability is 1 that only finitely many members of the sequence $\{S_n, n = 1, 2, \ldots\}$ are zero.

[*Hint*: Show that $\Pr(S_{2n} = 0) = \binom{2n}{n} p^n(1-p)^n$ and use Stirling's formula (2.5.15)]

6. Use the method Theorem 5.1.2 to write out a direct proof that if the B_n are independent, $P(\limsup_{n\to\infty} B_n) = 0$ implies $\Sigma\, P(B_n) < \infty$.

5.2 ALMOST SURE CONVERGENCE

If $\{u_n(x), n = 1, 2, \ldots\}$ is a sequence of functions of the real variable $x, x \in \mathscr{X}$, it is possible that the sequence converges to a limit function $u(x)$ for all or only some (or even no) $x \in \mathscr{X}$. For example, as $n \to \infty$,

$$u_n(x) = x + \sin nx, \qquad -\infty < x < \infty$$

converges to $u(x) = x$ when $x = 0, \pm\pi, \pm 2\pi, \ldots,$ but is oscillatory and does not converge for other $x \in (-\infty, \infty)$. This sequence converges only on a proper subset of the real line.

Similar limit behaviour can be exhibited by a sequence of random variables $\{X_n, n = 1, 2, \ldots\}$ defined on a probability space $(\Omega, \mathscr{F}, P)$. Each member of the sequence is a function from $\Omega = \{\omega\}$ into the real line and we may have convergence for all, some, or no $\omega \in \Omega$. Sets of the form $\{\omega \mid \lim_{n\to\infty} X_n(\omega) \text{ exists}\}$ and $\{\omega \mid \lim_{n\to\infty} X_n(\omega) = X(\omega)\}$ and the probabilities assigned them by P are of evident interest.

Example 1 $\Omega = \{\omega \mid \omega = -1, 0, 1\}$ and $X_n(\omega) = \omega^n, n = 1, 2, \ldots$. Then

$$X_n(1) = 1, \quad X_n(0) = 0, \quad X_n(-1) = (-1)^n.$$

Hence

$$\lim_{n\to\infty} X_n(1) = 1, \quad \lim_{n\to\infty} X_n(0) = 0,$$

but

$$\limsup_{n\to\infty} X_n(-1) = 1 \neq \liminf_{n\to\infty} X_n(-1) = -1$$

implying that $\lim_{n\to\infty} X_n(-1)$ does not exist. It follows that

$$\{\omega \mid \lim X_n(\omega) \text{ exists}\} = \{\omega \mid \omega = 0, 1\}$$

and its probability is

$$\Pr(\lim_{n\to\infty} X_n \text{ exists}) = P(\{0\}) + P(\{1\}). \quad \square$$

Example 2 $\Omega = \{\omega \mid = -1, 0, 1\}$ and

$$X_n(1) = 1, \quad X_n(0) = 0, \quad X_n(-1) = 1/n, \qquad n = 1, 2, \ldots$$

For all $\omega \in \Omega$ we have convergence to the random variable $X(\omega)$ defined by

$$X(1) = 1, \quad X(\omega) = 0 \qquad \text{if } \omega = -1, 0$$

and

$$Pr(\lim_{n\to\infty} X_n = X) = P(\{\omega \mid \lim X_n(\omega) = X(\omega)\}) = P(\Omega) = 1.$$

The distribution of the limit variable is

$$\Pr(X = 1) = P(\{1\}), \qquad \Pr(X = 0) = P(\{-1\}) + P(\{0\}). \quad \square$$

It is apparent that this mode of convergence is qualitatively different from the modes of convergence in probability and convergence in distribution considered in Chapter 4. These two latter treat the convergence of sequences of real numbers (which happen to be probabilities) or distribution functions, whereas our present interest centres on the set of sample points on which the random variables, considered as functions, converge to a limit function.

Definition 5.2.1 Suppose $\{X_n, n = 1, 2, \ldots\}$ is a sequence of random variables on $(\Omega, \mathscr{F}, P)$. If there is a random variable $X = X(\omega)$ on $(\Omega, \mathscr{F}, P)$ such that

$$P(\{\omega \mid \lim_{n\to\infty} X_n(\omega) = X(\omega)\}) = 1$$

then $\{X_n\}$ is said to converge *almost surely* (or with *probability* 1) to X, written

$$X_n \xrightarrow{\text{a.s.}} X. \quad \square$$

Two points should be noted. Firstly it is essential that all members of the sequence (and the limit, if it exists) be random variables on the same sample space, a requirement that is not necessary for convergence in distribution. Secondly, $X_n \xrightarrow{\text{a.s.}} X$ does not require pointwise convergence $\lim_{n\to\infty} X_n(\omega) = X(\omega)$ for all $\omega \in \Omega$, but only that the set of sample points for which $\lim_{n\to\infty} X_n(\omega) \neq X(\omega)$ is assigned probability zero by the probability function P.

The sequence $\{X_n(\omega), n = 1, 2, \ldots\}$ converges to the limit $X(\omega)$ at the sample point ω if the inequality $|X_n(\omega) - X(\omega)| < \varepsilon$ holds for all arbitrarily small $\varepsilon > 0$ and for all n sufficiently large. Then the set of sample points on which $X_n(\omega)$ does not converge to $X(\omega)$ is precisely that for which $|X_n(\omega) - X(\omega)| \geqslant \varepsilon$ for infinitely many n. It is convenient to have a more detailed description of this set of divergence.

Since ε is a small real number it is possible to find a positive integer m such that $(m+1)^{-1} < \varepsilon < m^{-1}$. m increases as ε decreases and the statement '$|X_n(\omega) - X(\omega)| \geqslant \varepsilon$ for arbitrarily small ε' is equivalent to '$|X_n(\omega) - X(\omega)| \geqslant m^{-1}$ for arbitrarily large m'. The set

$$B = \{\omega \,|\, X_n(\omega) - X(\omega)| \geqslant m^{-1} \text{ for infinitely many } n \text{ and arbitrarily large } m\}$$

is the set of sample points on which $X_n(\omega)$ does not converge to $X(\omega)$. Now for each $m = 1, 2, \ldots$, the sets

$$\begin{aligned} B_m &= \limsup_{n\to\infty} \{\omega \,|\, X_n(\omega) - X(\omega)| \geqslant m^{-1}\} \\ &= \bigcap_{n=1}^{\infty} \bigcup_{j=n}^{\infty} \{\omega \,|\, X_j(\omega) - X(\omega)| \geqslant m^{-1}\} \end{aligned} \tag{5.2.1}$$

are those for which the inequality $|X_n(\omega) - X(\omega)| \geqslant m^{-1}$ holds for infinitely many n. The set of divergence, B, is therefore

$$B = \bigcup_{m=1}^{\infty} B_m = \bigcup_{m=1}^{\infty} \bigcap_{n=1}^{\infty} \bigcup_{j=n}^{\infty} \{\omega \,|\, |X_j(\omega) - X(\omega)| \geqslant m^{-1}\}.$$

The complement of B is the set of sample points on which $X_n(\omega)$ converges to $X(\omega)$,

$$\bar{B} = \{\omega \,|\, \lim_{n\to\infty} X_n(\omega) = X(\omega)\},$$

and of course $P(\bar{B}) = 1 - P(B)$.

Theorem 5.2.1 $X_n \xrightarrow{\text{a.s.}} X$ if and only if $P(B_m) = 0$ for every $m \geqslant 1$, where the B_m are the sets defined in (5.2.1).

Proof. Suppose firstly that $X_n \xrightarrow{\text{a.s.}} X$. Then the divergence set $\bigcup_{m=1}^{\infty} B_m$ has probability zero. But $B_m \subseteq \bigcup_{j=1}^{\infty} B_j$ for $m = 1, 2, \ldots$, and hence $P(B_m) = 0$, $m \geqslant 1$. Conversely suppose $P(B_m) = 0$ for every $m \geqslant 1$. Then

$$P\left(\bigcup_{m=1}^{\infty} B_m\right) \leqslant \sum_{m=1}^{\infty} P(B_m) = 0,$$

implying that the complementary event $\{\omega \mid \lim_{n\to\infty} X_n(\omega) = X(\omega)\}$ has probability 1. □

The criterion for almost sure convergence provided by Theorem 5.2.1 is unfortunately not very helpful in practice. This is particularly true when information about members of the sequence comes in the form of specification of moments or distribution functions. However, a relatively simple sufficient condition which can frequently be restated in terms of distribution functions or their properties is the following:

Theorem 5.2.2 If $\sum_{n=1}^{\infty} \Pr(|X_n - X| \leqslant 1/m) < \infty$ for every $m \geqslant 1$ then $X_n \xrightarrow{\text{a.s.}} X$.

Proof. Since

$$\Pr(|X_n - X| \geqslant m^{-1}) = P(\{\omega \mid |X_n(\omega) - X(\omega)| \geqslant m^{-1}\})$$

by definition, the hypothesis can be stated as

$$\sum_{n=1}^{\infty} P(\{\omega \mid |X_n(\omega) - X(\omega)| \geqslant m^{-1}\}) < \infty \quad \text{for every } m \geqslant 1$$

Then part (i) of Theorem 5.1.2 implies

$$P(\limsup_{n\to\infty} \{\omega \mid |X_n(\omega) - X(\omega)| \geqslant m^{-1}\}) = 0$$

and an appeal to Theorem 5.2.1 completes the proof. □

Example 3 Let $\{X_n, n = 1, 2, \ldots\}$ be a sequence of random variables on a sample space with zero means and variances $\operatorname{Var} X_n$ satisfying the condition $\sum_{n=1}^{\infty} \operatorname{Var} X_n < \infty$. By Chebyshev's inequality (1.5.4)

$$\Pr(|X_n| \geqslant m^{-1}) \leqslant m^2 \operatorname{Var} X_n$$

for every $m \geqslant 1$. Consequently $\sum_{n=1}^{\infty} \Pr(|X_n| \geqslant m^{-1}) < \infty$, and by the Theorem $X_n \xrightarrow{\text{a.s.}} 0$. That is, this sequence of random variables converges almost surely to the random variable degenerate at the origin, $X(\omega) = 0$, all $\omega \in \Omega$. □

Example 4 It should be stressed that the condition of Theorem 5.2.2 is sufficient and generally not necessary. However it is relatively easy to verify, and is in fact necessary for certain classes of random variables. In particular, suppose $X_1, X_2, \ldots$ are independent with $X_n \xrightarrow{\text{a.s.}} 0$. Then

$$P(\limsup_{n \to \infty} \{\omega | \, |X_n(\omega)| \geqslant m^{-1}\}) = 0$$

no matter how large m is, and Exercise 5.1.6 (the assertion of which is essentially a restatement of Theorem 5.1.2(ii)) implies that

$$\sum_{n=1}^{\infty} \Pr(|X_n| \geqslant m^{-1}) < \infty.$$

For these independent variables the theorem is both necessary and sufficient. □

The condition* $\sum_{n=1}^{\infty} \Pr(|X_n - X| \geqslant m^{-1}) < \infty$, all $m \geqslant 1$, of Theorem 5.2.2 implies both $X_n \xrightarrow{\text{a.s.}} X$ and (since the general term of a convergent sum must tend to zero) the convergence in probability

* The sequence $\{X_n, n = 1, 2, \ldots\}$ is sometimes said to *converge completely* to X if $\sum_{n=1}^{\infty} \Pr(|X_n - X| \geqslant \varepsilon) < \infty$ for all $\varepsilon > 0$, but the term is not of wide currency. Complete convergence implies both almost sure convergence and convergence in probability but not conversely.

of X_n to X (see Definition 4.2.2). It is apparent that the two modes of convergence are distinct and it is natural to ask in what way they are related.

Theorem 5.2.3 $X_n \xrightarrow{\text{a.s.}} X$ implies $X_n \xrightarrow{\text{p}} X$.

Proof. The sets B_m in (5.2.1) can be written (see (5.1.1)) as the limits

$$B_m = \lim_{n\to\infty} \bigcup_{j=n}^{\infty} \{\omega \mid |X_j(\omega) - X(\omega)| \geqslant m^{-1}\}$$

with (from (5.1.2)) probabilities

$$P(B_m) = \lim_{n\to\infty} P\Big(\bigcup_{j=n}^{\infty} \{\omega \mid |X_j(\omega) - X(\omega)| \geqslant m^{-1}\}\Big).$$

Now for each $n = 1, 2, \ldots,$

$$\begin{aligned}\Pr(|X_n - X| \geqslant m^{-1}) &= P(\{\omega \mid |X_n(\omega) - X(\omega)| \geqslant m^{-1}\}) \\ &\leqslant P\Big(\bigcup_{j=n}^{\infty} \{\omega \mid |X_j(\omega) - X(\omega)| \geqslant m^{-1}\}\Big)\end{aligned}$$

and hence

$$\lim_{n\to\infty} \Pr(|X_n - X| \geqslant m^{-1}) \leqslant P(B_m).$$

However, by Theorem 5.2.1, $P(B_m) = 0$ for every $m \geqslant 1$ if $X_n \xrightarrow{\text{a.s.}} X$. Therefore $\Pr(|X_n - X| \geqslant m^{-1}) \to 0$ as $n \to \infty$ for all $m \geqslant 1$, which is precisely the condition for $X_n \xrightarrow{\text{p}} X$. □

The converse proposition that $X_n \xrightarrow{\text{p}} X$ implies $X_n \xrightarrow{\text{a.s.}} X$ is not true except for certain special sequences (see Exercises 4 and 5). Indeed for arbitrarily small $\varepsilon > 0$, $\delta > 0$, and large $N = N(\varepsilon, \delta)$, $M = M(\varepsilon, \delta)$, the definitions of $X_n \xrightarrow{\text{a.s.}} X$ and $X_n \xrightarrow{\text{p}} X$ can be restated respectively as

$$P\Big(\bigcap_{n=N}^{\infty} \{\omega \mid |X_n(\omega) - X(\omega)| < \varepsilon\}\Big) > 1 - \delta$$

and

$$P(\{\omega \mid |X_n(\omega) - X(\omega)| < \varepsilon\}) > 1 - \delta \qquad \text{for } n \geqslant M.$$

Almost sure convergence requires that the *simultaneous occurrence* of the events

$$\{\omega \,|\, |X_n(\omega) - X(\omega)| < \varepsilon\}, \qquad n = N, N+1, \ldots,$$

has probability close to 1, which is not the case for convergence in probability. Recalling Theorem 4.2.3 we see that the three modes of convergence considered so far are related in the following way;

Almost sure convergence ⇒ Convergence in probability
⇒ Convergence in distribution,

with reverse implications holding only in special cases.

Exercises

1. The random variables $X_1, X_2, X_3, \ldots$, are defined on

$$\Omega = \{\omega \,|\, \omega = \pm 1, \pm 2\}$$

by

$$\left.\begin{array}{l} X_n(-1) = -1, \quad X_n(1) = 1 \\ X_n(-2) = 0 = X_n(2) \end{array}\right\} \text{ if } n \text{ odd,}$$

$$\left.\begin{array}{l} X_n(-1) = 0 = X_n(1) \\ X_n(-2) = -2, \quad X_n(2) = 2 \end{array}\right\} \text{ if } n \text{ even.}$$

Find the distributions of the random variables $^*X = \limsup_{n\to\infty} X_n$, $_*X = \liminf_{n\to\infty} X_n$, and show that $\Pr(\lim_{n\to\infty} X_n \text{ exists}) = 0$.

2. $\{C_n, n = 1, 2, \ldots\}$ is a sequence of events on $(\Omega, \mathscr{F}, P)$ and $\{I_n, n = 1, 2, \ldots\}$ is the sequence of corresponding indicator variables, i.e. $I_n(\omega) = 1$ if $\omega \in C_n$ and is zero otherwise. Show

(a) $\sum_{n=1}^{\infty} P(C_n) < \infty$ implies the almost sure convergence of $\sum_{n=1}^{\infty} I_n$,

(b) if the C_n are independent and $\sum_{n=1}^{\infty} P(C_n) = \infty$ then $\sum_{n=1}^{\infty} I_n = \infty$ almost surely.

3. Use the result of the preceding exercise to show that in an infinite sequence of Bernoulli trials the probability is one that the pattern 'four successive heads' occurs infinitely many times.

4. $X_1, X_2, X_3, \ldots,$ are independent random variables with distributions

$$\Pr(X_n = 0) = \frac{1}{n}, \qquad \Pr(X_n = 1) = 1 - \frac{1}{n}.$$

Clearly $X_n \xrightarrow{\text{P}} 1$. However X_n does not converge to 1 almost surely. Show this by applying the second Borel–Cantelli Lemma, Theorem 5.1.2 (ii), to the events $\{\omega \,|\, X_n(\omega) = 0\}, n = 1, 2, \ldots$.

5. Suppose $\{X_n, n = 1, 2, \ldots\}$ is a monotone sequence of non-negative random variables satisfying $X_1(\omega) \geqslant X_2(\omega) \geqslant X_3(\omega) \geqslant \ldots,$ all $\omega \in \Omega$. Show that if $X_n \xrightarrow{\text{P}} 0$ then also $X_n \xrightarrow{\text{a.s.}} 0$. We have here a a sequence in which the two modes of convergence are equivalent. [*Hint*: Show that monotonicity implies

$$\bigcup_{j=n}^{\infty} \{\omega \,|\, X_j(\omega) \geqslant m^{-1}\} = \{\omega \,|\, X_n(\omega) \geqslant m^{-1}\}$$

for every $n \geqslant 1$, and hence $P(B_m) = \lim_{n \to \infty} P(\{\omega \,|\, X_n(\omega) \geqslant m^{-1}\})$].

6. *Cauchy criterion* Show that $\{X_n\}$ converges almost surely if and only if $|X_n - X_m| \xrightarrow{\text{a.s.}} 0$ as $n, m \to \infty$.

7. It is true that if $X_1, X_2, \ldots,$ are independent and $\sum_{j=1}^{n} X_j$ converges in distribution as $n \to \infty$, then $\sum_{j=1}^{n} X_j$ also converges almost surely and in probability. Omit the proof but verify the result for independent summands X_n with

$$\Pr(X_n = 1) = p^n, \quad \Pr(X_n = 0) = 1 - p^n, \qquad 0 < p < 1.$$

For these partial sums the three modes of convergence are equivalent (c.f. Theorem 4.2.4).

5.3 THE STRONG LAW OF LARGE NUMBERS

The classical form of the weak law of large numbers (§ 4.2) stated that if S_n is the sum of n independently and identically distributed variables with common expectation μ than for arbitrary $\varepsilon > 0$.

$$\lim_{n\to\infty} \Pr(|n^{-1}S_n - \mu| < \varepsilon) = 1. \tag{5.3.1}$$

That is, the sequence of random variables $\{n^{-1}S_n\}$ converges in probability to the constant μ, $n^{-1}S_n \xrightarrow{P} \mu$.

On the other hand the strong law of large numbers for identically distributed variables states that

$$\Pr(\lim_{n\to\infty} n^{-1}S_n = \mu) = 1. \tag{5.3.2}$$

The strong law asserts the almost sure convergence of $\{n^{-1}S_n\}$ to μ, i.e. that the probability is 1 that $|n^{-1}S_n - \mu| < \varepsilon$ for infinitely many values of n. In keeping with remarks in the introduction to this chapter we assume the existence of a probability space $(\Omega, \mathscr{F}, P)$ on which all members of the sequence $\{S_n, n = 1, 2, \ldots\}$ are defined, and it is in terms of this probability space that (5.3.2) translates as

$$P(\{\omega \mid \lim_{n\to\infty} n^{-1}S_n(\omega) = \mu\}) = 1.$$

In this section we prove (5.3.2) when a moment generating function exists and the more general result is established in § 5.4. Thus consider the sequence

$$S_n = Y_1 + Y_2 + \ldots + Y_n, \qquad n = 1, 2, \ldots,$$

with the Y_j independently identically distributed as Y, and Ee^{-sY} defined for s in a non-degenerate interval including the origin. This implies, in particular that

$$\mu = -\frac{d}{ds}(Ee^{-sY})\bigg|_{s=0}$$

is finite.

The lemma on which the proof depends is of some independent interest and is an extension of Chebyshev's inequality (1.5.4).

Lemma Suppose $Ee^{-sY} = M(s)$ is defined for at least $-h < s < h$. Then there exists an $\alpha, 0 < \alpha < 1$, such that for any $\varepsilon > 0$

$$\Pr(|n^{-1}S_n - \mu| \geqslant \varepsilon) \leqslant 2\alpha^n.$$

Proof.

$$\Pr(|n^{-1}S_n - \mu| \geqslant \varepsilon) = \Pr(S_n \geqslant n(\mu+\varepsilon)) + \Pr(S_n \leqslant n(\mu-\varepsilon)).$$

We find bounds for the two probabilities on the right hand side. Consider first $\Pr(S_n \leqslant n(\mu-\varepsilon))$ and let

$$Z = Y - \mu + \varepsilon.$$

Then

$$Ee^{-sZ} = Ee^{-s(Y-\mu+\varepsilon)} = e^{s(\mu-\varepsilon)}M(s), \qquad -h < s < h,$$

and

$$EZ = \varepsilon = -\frac{d}{ds}(Ee^{-sZ})\bigg|_{s=0} > 0$$

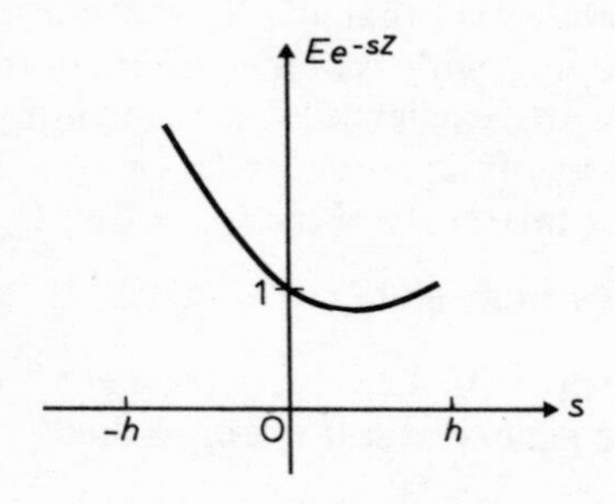

The moment generating function Ee^{-sZ} therefore has negative slope at the origin and hence there exists an s_0 in $(0, h)$ such that

$$Ee^{-s_0 Z} < 1.$$

Let $Z_1, Z_2, \ldots, Z_n$ be independently and identically distributed as Z, and let $f_n(z)$ be the density function of

$$\sum_{j=1}^{n} Z_j = S_n - n(\mu-\varepsilon).$$

Recognizing that the moment generating function of $\sum_{j=1}^{n} Z_j$ is the

nth power of Ee^{-sZ}, we have

$$\begin{aligned}\{E\exp(-s_0Z)\}^n &= E\{\exp(-s_0\sum_{j=1}^{n} Z_j)\} \\ &= \int_{-\infty}^{\infty} e^{-s_0z} f_n(z)\,dz \\ &\geqslant \int_{-\infty}^{0} e^{-s_0z} f_n(z)\,dz \\ &\geqslant \int_{-\infty}^{0} f_n(z)\,dz, \qquad \text{since } s_0 > 0, \\ &= \Pr(\sum_{j=1}^{n} Z_j \leqslant 0) = \Pr(S_n \leqslant n(\mu-\varepsilon)).\end{aligned}$$

Writing $\alpha_0 = Ee^{-s_0Z}$, it follows that

$$\Pr(S_n \leqslant n(\mu-\varepsilon)) \leqslant \alpha_0^n \qquad \text{with } 0 < \alpha_0 < 1.$$

In precisely the same way we can introduce

$$W = Y-\mu-\varepsilon$$

and find a number $s_1 < 0$ such that $\alpha_1 = Ee^{-s_1W} < 1$ and

$$\Pr(S_n \geqslant n(\mu+\varepsilon)) \leqslant \alpha_1^n.$$

If $\alpha = \max(\alpha_0, \alpha_1)$, the inequality

$$\Pr(S_n \geqslant n(\mu+\varepsilon)) + \Pr(S_n \leqslant n(\mu-\varepsilon)) \leqslant 2\alpha^n$$

follows, and the lemma is proved. □

Theorem 5.3.1 If $Y_1, Y_2, \ldots$ are independently and identically distributed as Y, whose moment generating function Ee^{-sY} is defined in a non-degenerate interval containing the origin, then

$$\Pr(\lim_{n\to\infty} n^{-1}S_n = \mu) = 1,$$

where $S_n = Y_1, + Y_2 + \ldots + Y_n$ and $\mu = EY$.

Proof. By the Lemma an α, $0 < \alpha < 1$, can be found such that for any $\varepsilon > 0$

$$\Pr(|n^{-1}S_n - \mu| \geqslant \varepsilon) \leqslant 2\alpha^n.$$

Hence

$$\sum_{n=1}^{\infty} \Pr(|n^{-1}S_n - \mu| \geqslant \varepsilon) \leqslant 2 \sum_{n=1}^{\infty} \alpha^n = 2\alpha/(1-\alpha) < \infty.$$

By Theorem 5.2.2 this implies $n^{-1}S_n \xrightarrow{\text{a.s.}} \mu$, i.e. $\Pr(\lim_{n\to\infty} n^{-1}S_n = \mu) = 1$.

□

Example 1 Bernoulli trials with $\Pr(Y = 1) = p$, $\Pr(Y = 0) = 1-p$. The moment generating function is

$$Ee^{-sY} = 1-p+pe^{-s}$$

which exists for all finite s. The theorem applies and

$$\Pr(\lim_{n\to\infty} n^{-1}S_n = p) = 1. \qquad \square$$

It is Theorem 5.3.1 and Example 1 that provide the basis for the 'frequency' interpretation of probability. Let A be an event on the probability space $(\Omega', \mathscr{F}', P')$ and denote its probability by

$$p = P'(A).$$

Suppose it is possible to carry out n independent and identical experiments at each of which we can determine whether or not A has occurred. Defining the indicator variables $I_A^{(1)}, I_A^{(2)}, \ldots,$ by

$$\begin{aligned} I_A^{(j)} &= 1 \quad \text{if } A \text{ occurs at the } j\text{th experiment,} \\ &= 0 \quad \text{if } A \text{ does not occur at the } j\text{th experiment,} \end{aligned}$$

the sum

$$S_n = I_A^{(1)} + I_A^{(2)} + \ldots + I_A^{(n)}$$

of these independent and identically distributed indicator variables is the number of times A occurs in the first n experiments. If $\Omega_1 = \{0, 1\}$ is the sample space of $I_A^{(1)}$ then S_n is a random variable on

$$\Omega^{(n)} = \Omega_1 \times \Omega_1 \times \ldots \times \Omega_1, \qquad n \text{ times,}$$

and its distribution for each fixed n is

$$\Pr(S_n = j) = \binom{n}{j} p^j (1-p)^{n-j}, \qquad j = 0, 1, \ldots, n. \qquad (5.3.3)$$

As n becomes arbitrarily large, $\Omega^{(n)} \to \Omega$, where Ω is the sample space whose points are vectors of infinite length, each coordinate being 0 or 1. By the Extension Theorem we can assign in a unique way a probability function P on Ω that is consistent with (5.3.3). Then Example 1 applies and

$$n^{-1}S_n \overset{\text{a.s.}}{\to} p.$$

Now the ratio $n^{-1}S_n$ is the relative frequency with which A occurs in n experiments. The event $\{\omega \mid \lim_{n\to\infty} n^{-1}S_n = p\}$ is the set of sample points (infinite vectors) on which this relative frequency is arbitrarily close to p for an arbitrarily large number of experiments. According to Theorem 5.3.1 this event has probability 1. The set of sample points for which $n^{-1}S_n$ does not converge to p is negligible in the sense of having probability zero. Thus within our formal framework we have for any observable event A a frequency interpretation of its probability p as the almost sure limit of its relative frequency of occurrence.

The frequency interpretation should be compared with the subjective one mentioned briefly in the course of the discussion on Bayes's Theorem in § 3.4. The two points of view are by no means necessarily incompatible although a frequentist tends to discount the use of Bayes's Theorem as a serious form of inductive inference except when the prior distribution can be prescribed in an 'objective' way.

Returning to the proof of Theorem 5.3.1 note that the method consists of finding a sequence $\{a_n\}$ dominating the terms

$$\Pr(|n^{-1}S_n - \mu| \geqslant \varepsilon)$$

and such that $\sum a_n < \infty$. Assuming a moment generating function we found $a_n = 2\alpha^n$, $0 < \alpha < 1$. The method still works under less restrictive conditions. For example, in Exercise 2 you are asked to show that if $EY^4 < \infty$ then $n^{-1}S_n \to \mu$. However use of Theorem 5.2.2 yields only a sufficient condition for almost sure convergence and the complete form of the strong law (Theorem 5.4.5) requires a more sophisticated technique for its proof.

Finally we note that it is possible for very general sequences of random variables to obey a strong law. Indeed if $S_n = X_1 + X_2 + \ldots + X_n$ and $n^{-1}\{S_n - ES_n\} \overset{\text{a.s.}}{\to} 0$ then the strong law of large numbers is said to hold for $\{X_n\}$. Examples appear in the Exercises and in the next section.

Exercises

1. The density function of Y is $f(y) = \lambda e^{-\lambda y}$, $y > 0$, and zero otherwise. For $0 < \varepsilon < \lambda^{-1}$ show that the sharpest inequality given by the argument of the lemma is

$$\Pr(|n^{-1}S_n - \lambda^{-1}| \geqslant \varepsilon) \leqslant ((1-\lambda\varepsilon)e^{\lambda\varepsilon})^n + ((1+\lambda\varepsilon)e^{-\lambda\varepsilon})^n$$

Deduce directly that the negative exponential distribution obeys the strong law of large numbers. Obtain the corresponding inequality if $Y \frown N(\mu, \sigma^2)$.

2. Show that if $EY = 0$ and $EY^4 < \infty$ then there is a positive constant C such that $E(n^{-1}S_n)^4 \leqslant Cn^{-2}$. Use this inequality to establish the strong law of large numbers for random variables with finite fourth moment.

3. The proof of the lemma on which Theorem 5.3.1 depends was phrased in terms of continuous variables and only trivial changes are required to adapt to the discrete case. However, supposing that Y is distributed over the integers, prove directly that if Ez^Y converges for $z \in (1-h_1, 1+h_1)$ then $\Pr(|n^{-1}S_n - \mu| \geqslant \varepsilon) \leqslant 2\beta^n$ for some $\beta \in (0, 1)$. Verify your argument by considering the case

$$\Pr(Y = 1) = p, \Pr(Y = 0) = 1-p.$$

4. Prove that the strong law of large numbers holds for independent bounded random variables.
[*Hint*: Show that any bounded random variable has a moment generating function.]

5. $X_1, X_2, \ldots$ are independent with

$$\Pr\left(X_n = 1+\frac{1}{n}\right) = \frac{1}{2}\left\{1+\left(1-\frac{1}{n^2}\right)^{\frac{1}{2}}\right\}$$

$$\Pr\left(X_n = -1-\frac{1}{n}\right) = \frac{1}{2}\left\{1-\left(1-\frac{1}{n^2}\right)^{\frac{1}{2}}\right\}.$$

Show that the strong law holds.

6. Suppose X_j may depend on X_{j-1} but is independent of other members of the sequence $\{X_n, n = 1, 2, \ldots\}$. Show that the weak

law holds if the variances are uniformly bounded. Do both laws of large numbers hold if for some $\delta > 0$, $\text{Var}X_n \sim n^{-\delta}$ as $n \to \infty$?

5.4* THE STRONG LAW (CONTINUED)

The purpose of this section is to show that the existence of a first moment is necessary and sufficient for a sequence of independent and identically distributed variables to obey the strong law of large numbers. Recall (Theorems 4.2.1,2) that a finite mean is sufficient but not necessary for the weak law to hold.

The proof requires several preliminary results that are of wide application and of sufficient importance to be stated as theorems in their own right. The first of these is an inequality that is considerably stronger than Chebyshev's, to which it reduces when $n = 1$.

Theorem 5.4.1 (Kolmogorov's Inequality) Let $X_1, X_2, \ldots,$ be independent random variables with zero means and finite variances. If $S_j = X_1 + X_2 + \ldots + X_j, j = 1, 2, \ldots, n$, then for any $\varepsilon > 0$

$$\Pr(\max\{|S_1|, |S_2|, \ldots, |S_n|\} \geqslant \varepsilon) \leqslant \frac{1}{\varepsilon^2} \sum_{j=1}^{n} \text{Var}\, X_j. \tag{5.4.1}$$

Proof. Introduce the events

$$B_1 = \{\omega \,|\, |S_1| \geqslant \varepsilon\}$$
$$B_j = \{\omega \,|\, |S_i| < \varepsilon, i = 1, 2, \ldots, j-1, \text{ and } |S_j| \geqslant \varepsilon\}, \quad j = 2, 3, \ldots, n,$$

and

$$B = \{\omega \,|\, \max(|S_1|, |S_2|, \ldots, |S_n|) \geqslant \varepsilon\}.$$

B_j is the event that the first maximum is achieved by the jth partial sum which has a value $\geqslant \varepsilon$. Clearly the B_j are disjoint and

$$B = \bigcup_{j=1}^{n} B_j.$$

Let I_B, I_{B_j} respectively denote the indicator variables of B and B_j, $j = 1, 2, \ldots, n$. Then

$$E(I_{B_j} S_n^2) = E[I_{B_j}\{S_j + (S_n - S_j)\}^2]$$
$$= E(I_{B_j} S_j^2) + 2E\{I_{B_j} S_j(S_n - S_j)\} + E\{I_{B_j}(S_n - S_j)^2\}.$$

* To be omitted at a first reading.

The X_k are independent and $S_n - S_j$ depends only on $X_{j+1}, X_{j+2}, \dots, X_n$. Hence

$$\begin{aligned} E\{I_{B_j} S_j(S_n - S_j)\} &= \{E(I_{B_j} S_j)\}\{E(S_n - S_j)\} \\ &= 0, \quad \text{since } EX_k = 0, k = 1, 2 \dots, n. \end{aligned}$$

Also

$$\begin{aligned} E\{I_{B_j}(S_n - S_j)^2\} &= \{EI_{B_j}\}\{E(S_n - S_j)^2\} \\ &\geqslant 0. \end{aligned}$$

But $I_{B_j} S_j^2$ is zero if B_j does not occur and is $\geqslant \varepsilon^2$ if B_j does occur. Therefore

$$E(I_{B_j} S_j^2) \geqslant \varepsilon^2 \Pr(I_{B_j} = 1)$$

and

$$\begin{aligned} E(I_{B_j} S_n^2) &= E(I_{B_j} S_j^2) + E\{(I_{B_j}(S_n - S_j)^2\} \\ &\geqslant \varepsilon^2 \Pr(I_{Bj} = 1). \end{aligned} \tag{5.4.2}$$

Now B is the union of the disjoint events $B_1, B_2, \dots, B_n$ which implies that

$$\sum_{j=1}^{n} I_{B_j} = I_B$$

and

$$\sum_{j=1}^{n} \Pr(I_{B_j} = 1) = \sum_{j=1}^{n} P(B_j) = P(B).$$

Summing the inequalities (5.4.2) for $j = 1, 2, \dots n$, yields

$$\begin{aligned} \varepsilon^2 P(B) &= \varepsilon^2 \sum_{j=1}^{n} \Pr(I_{B_j} = 1) \\ &\leqslant \sum_{j=1}^{n} E(I_{B_j} S_n^2) \\ &= E(I_B S_n^2) \\ &\leqslant ES_n^2 = \sum_{j=1}^{n} \operatorname{Var} X_j, \end{aligned}$$

the final inequality holding because $I_B S_n^2 \leqslant S_n^2$. Recalling the definition of B we see that the proof of (5.4.1) is complete. □

Kolmogorov's inequality does not require the summands to be identically distributed and indeed the important applications from our point of view are to sums of independent but not identically distributed variables. One immediate consequence is

Theorem 5.4.2 If $X_1, X_2, \ldots,$ are independent random variables with zero means and whose variances are such that $\sum_{n=1}^{\infty} \operatorname{Var} X_n < \infty$, then

$$\Pr\left(\sum_{n=1}^{\infty} X_n \text{ converges}\right) = 1.$$

Proof. Since $S_{N+n} - S_n = X_{n+1} + X_{n+2} + \ldots + X_{N+n}$ it follows from (5.4.1) that for any $\varepsilon > 0$

$$\Pr\left(\max_{1 \leqslant N \leqslant k} \{|S_{N+n} - S_n|\} \geqslant \varepsilon\right) \leqslant \frac{1}{\varepsilon^2} \sum_{j=n+1}^{n+k} \operatorname{Var} X_j.$$

Letting $k \to \infty$ this becomes

$$\Pr\left(\max_{1 \leqslant N} \{|S_{N+n} - S_n|\} \geqslant \varepsilon\right) = \Pr\left(\max_{n<m} \{|S_m - S_n|\} \geqslant \varepsilon\right) \leqslant \frac{1}{\varepsilon^2} \sum_{j=n+1}^{\infty} \operatorname{Var} X_j.$$

But $\lim_{n \to \infty} \sum_{j=n+1}^{\infty} \operatorname{Var} X_j = 0$, since $\sum \operatorname{Var} X_j < \infty$, which implies for the complementary event

$$\lim_{n \to \infty} \Pr\left(\max_{n<m} \{|S_m - S_n|\} < \varepsilon\right) = 1.$$

Since $\varepsilon > 0$ is arbitrary this in turn implies

$$\Pr\left(\lim_{n \to \infty} \left|\sum_{j=n}^{\infty} X_j\right| = 0\right) = 1$$

and consequently $\sum_{j=1}^{\infty} X_j$ converges almost surely. □

The next theorem is a version of the strong law for non-identically distributed variables.

Theorem 5.4.3 Let $X_1, X_2, \ldots,$ be independent random variables with means $\mu_n = EX_n$. If

$$\sum_{n=1}^{\infty} \frac{1}{n^2} \operatorname{Var} X_n < \infty$$

R

then

$$\Pr\left(\lim_{n\to\infty}\left\{\frac{1}{n}\sum_{j=1}^{n} X_j - \frac{1}{n}\sum_{j=1}^{n} \mu_j\right\} = 0\right) = 1.$$

Proof. The random variables

$$X_n^* = n^{-1}(X_n - \mu_n)$$

have zero means and variances $\operatorname{Var} X_n^* = (\operatorname{Var} X_n)n^{-2}$. Theorem 5.4.2 applies to yield

$$\Pr(\sum_{n=1}^{\infty} n^{-1}(X_n - \mu_n) \text{ converges}) = 1.$$

That the almost sure convergence of $\sum_{n=1}^{\infty} n^{-1}(X_n - \mu_n)$ implies the desired result can be seen as follows. Put

$$Z_n = \sum_{j=1}^{n} j^{-1}(X_j - \mu_j).$$

Then $n(Z_n - Z_{n-1}) = X_n - \mu_n, n = 1, 2, \ldots,$ and

$$\begin{aligned} n^{-1}\sum_{j=1}^{n}(X_j - \mu_j) &= n^{-1}\{Z_1 + 2(Z_2 - Z_1) + 3(Z_3 - Z_2) \\ &\quad + \ldots + n(Z_n - Z_{n-1})\} \\ &= Z_n - n^{-1}\sum_{j=1}^{n-1} Z_j. \end{aligned}$$

But Z_n converges (a.s.) to Z, say, and (Exercise 2.2.11) the average $n^{-1}\sum_{j=1}^{n-1} Z_j$ also converges (a.s.) to Z. Hence as $n \to \infty$

$$n^{-1}\sum_{j=1}^{n}(X_j - \mu_j) \overset{\text{a.s.}}{\to} 0$$

which was to be proved. □

A corollary is that if the independent $X_1, X_2, \ldots,$ are identically distributed with mean μ and finite variance then $n^{-1}\sum_{j=1}^{n} X_j \overset{\text{a.s.}}{\to} \mu$. Our aim is to dispense with assumptions about moments higher than the first. The following two lemmas give necessary and sufficient conditions for a first moment in forms that will subsequently prove useful. Recall that for any random variable X, EX exists if and only if $E|X| < \infty$.

Lemma 1 $E|X| < \infty$ if and only if $\sum \Pr(|X| \geqslant n) < \infty$.

Proof. The lemma is an immediate consequence of the integral test for the convergence of the sum of non-increasing terms. For (Theorem 1.5.1)

$$EX = \int_0^\infty \Pr(X > x)\, dx - \int_{-\infty}^0 \Pr(X \leqslant x)\, dx$$

and the two integrals on the right hand side converge or diverge with $\sum_{n=0}^{\infty} \Pr(X \geqslant n)$ and $\sum_{n=0}^{\infty} \Pr(X \leqslant -n)$ respectively. □

Lemma 2 $E|X| < \infty$ if and only if $\sum_{n=1}^{\infty} \frac{1}{n^2} \int_{-n}^{n} x^2 f(x)\, dx < \infty$.

Proof.

$$\begin{aligned}
&\sum_{n=1}^{\infty} \frac{1}{n^2} \int_{-n}^{n} x^2 f(x)\, dx \\
&\qquad = \sum_{n=1}^{\infty} \frac{1}{n^2} \sum_{j=0}^{n-1} \left\{ \int_j^{j+1} x^2 f(x)\, dx + \int_{-j-1}^{-j} x^2 f(x)\, dx \right\} \\
&\qquad = \sum_{j=0}^{\infty} \left\{ \int_j^{j+1} x^2 f(x)\, dx + \int_{-j-1}^{-j} x^2 f(x)\, dx \right\} \sum_{n=j+1}^{\infty} 1/n^2.
\end{aligned}$$

But for $j > 0$ there are positive constants c_1 and c_2 such that

$$\frac{c_1}{j} \leqslant \sum_{n=j+1}^{\infty} \frac{1}{n^2} \leqslant \frac{c_2}{j+1}.$$

Also

$$\begin{aligned}
&j\left\{ \int_j^{j+1} x f(x)\, dx + \int_{-j-1}^{-j} |x| f(x)\, dx \right\} \\
&\qquad \leqslant \int_j^{j+1} x^2 f(x)\, dx + \int_{-j-1}^{-j} x^2 f(x)\, dx \\
&\qquad \leqslant (j+1)\left\{ \int_j^{j+1} x\, f(x)\, dx + \int_{-j-1}^{-j} |x| f(x)\, dx \right\}.
\end{aligned}$$

Since

$$E|X| = \int_0^\infty x\, f(x)\, dx + \int_{-\infty}^0 |x| f(x)\, dx$$
$$= \sum_{j=0}^\infty \left\{ \int_j^{j+1} x\, f(x)\, dx + \int_{-j-1}^{-j} |x| f(x)\, dx \right\}.$$

it follows that

$$c_1 E|X| \leqslant \sum_{n=1}^\infty \frac{1}{n^2} \int_{-n}^n x^2 f(x)\, dx \leqslant \{1 + c_2 E|X|\}.$$

Hence $E|X|$ and $\sum n^{-2} \int_{-n}^n x^2 f(x)\, dx$ are finite or infinite together, which is what we want to prove. □

We now introduce an important device which will lead to the proof of our main result. With any sequence of random variables $\{Y_n, n = 1, 2, \ldots\}$ we can associate a sequence $\{T_n, n = 1, 2, \ldots\}$ of *truncated variables* defined by

$$T_n = \begin{cases} Y_n & \text{if } |Y_n| < n \\ 0 & \text{if } |Y_n| \geqslant n. \end{cases} \tag{5.4.3}$$

Truncated variables are often easier to handle than the original ones, and furthermore, under wide conditions, the two sequences have the same asymptotic behaviour.

Theorem 5.4.4 Let $\{Y_n\}$ be a sequence of independent random variables identically distributed as Y, and let $\{T_n\}$ be the associated truncated variables as in (5.4.3). Then, if $E|Y| < \infty$, the probability that $T_n \neq Y_n$ for infinitely many n is zero. Consequently if

$$n^{-1} \sum_{j=1}^n T_j \xrightarrow{\text{a.s.}} C \quad \text{then} \quad n^{-1} \sum_{j=1}^n Y_j \xrightarrow{\text{a.s.}} C$$

and conversely.

Proof. For $n = 1, 2, 3, \ldots$, let

$$B_n = \{\omega \,|\, T_n \neq Y_n\}.$$

We want to show that $E|Y| < \infty$ implies $P(\limsup_{n\to\infty} B_n) = 0$.

However

$$\limsup_{n\to\infty} B_n = \bigcap_{n=1}^{\infty}\bigcup_{j=n}^{\infty} B_j \subset \bigcup_{j=n}^{\infty} B_j$$

and hence

$$P(\limsup_{n\to\infty} B_n) \leqslant \sum_{j=n}^{\infty} P(B_j) = \sum_{j=n}^{\infty} \Pr(|Y| \geqslant j).$$

By Lemma 1 the finiteness of $E|Y|$ is equivalent to

$$\lim_{n\to\infty} \sum_{j=n}^{\infty} \Pr(|Y| \geqslant j) = 0,$$

which proves the required result. Now suppose A is the event on which $n^{-1}\sum_1^n T_j$ and $n^{-1}\sum_1^n Y_j$ converge (a.s.) to different limits. Then

$$A \subset \limsup_{n\to\infty} B_n.$$

Consequently $E|Y| < \infty$ implies $P(A) = 0$, and the complementary event on which the two sequences converge to the same limit (or diverge together) must have probability 1. □

Sequences such as the $\{T_n\}$ and $\{Y_n\}$ of this theorem are sometimes said to be *asymptotically equivalent*, meaning that the probability is one that they differ in only a finite number of terms. Our method of proving the strong law of large numbers is to first establish it for the sequence of truncated variables $\{T_n\}$ and then apply Theorem 5.4.4.

Theorem 5.4.5 (Strong law of large numbers) Let

$$S_n = Y_1 + Y_2 + \ldots + Y_n,$$

where the Y_j are independently and identically distributed as Y. The necessary and sufficient condition for

$$n^{-1} S_n \overset{\text{a.s.}}{\to} \mu$$

is that EY exist and be equal to μ.

Proof. Suppose firstly that $EY = \mu$ exists and consider the sequence $\{T_n\}$ of (5.4.3). For each finite n, $\operatorname{Var} T_n$ is finite and in fact

$$\operatorname{Var} T_n \leqslant ET_n^2 = \int_{-n}^{n} y^2 f(y)\, dy,$$

where $f(y)$ is the density of Y. Then by Lemma 2, $E|Y| < \infty$ implies

$$\sum_{n=1}^{\infty} \frac{1}{n^2} \operatorname{Var} T_n \leqslant \sum_{n=1}^{\infty} \frac{1}{n^2} \int_{-n}^{n} y^2 f(y)\, dy < \infty,$$

and, by Theorem 5.4.3,

$$\frac{1}{n} \sum_{j=1}^{n} T_j - \frac{1}{n} \sum_{j=1}^{n} ET_j \overset{\text{a.s.}}{\rightarrow} 0.$$

But as $n \rightarrow \infty$

$$ET_n = \int_{-n}^{n} y f(y)\, dy \rightarrow \mu.$$

Consequently $n^{-1} \sum_{1}^{n} T_j \overset{\text{a.s.}}{\rightarrow} \mu$, and from Theorem 5.4.4,

$$n^{-1} \sum_{1}^{n} Y_j \overset{\text{a.s.}}{\rightarrow} \mu.$$

To prove necessity we first show that $n^{-1} \sum_{1}^{n} Y_j \overset{\text{a.s.}}{\rightarrow} \mu$, with μ some finite number, implies $E|Y| < \infty$. Now the finiteness (a.s.) of $\lim_{n\rightarrow\infty} n^{-1} \sum_{1}^{n} Y_j$ ensures that

$$n^{-1} Y_n \overset{\text{a.s.}}{\rightarrow} 0.$$

Hence the probability is 0 that infinitely many of the inequalities $|Y_n| \geqslant n$ occur, or equivalently that

$$P(\limsup_{n\rightarrow\infty} \{\omega \,|\, |Y_n| \geqslant n\}) = 0.$$

Appealing to Theorem 5.1.2 (ii) (see also Exercise 5.1.6) we therefore have, since the Y_n are independent and identically distributed,

$$\infty > \sum_{n=1}^{\infty} \Pr(|Y_n| \geqslant n) = \sum_{n=1}^{\infty} \Pr(|Y| \geqslant n).$$

The existence of EY follows from Lemma 1, and Theorem 5.4.4 applies to yield that the truncated sequence $\{T_n\}$ and $\{Y_n\}$ are asymptotically equivalent. But the sufficiency part of the proof showed that $n^{-1} \sum_{1} T_j \overset{\text{a.s.}}{\rightarrow} EY$ and hence we must have $EY = \mu$. □

The strong law is thus essentially a first moment theorem. Indeed if EY does not exist, so that $E|Y| = \infty$, Lemma 1 implies

$$\sum_{n=1}^{\infty} \Pr(|Y_n| \geqslant n) = \infty.$$

Then, from the second Borel–Cantelli Lemma (Theorem 5.1.2. (ii)), the probability is 1 that infinitely many of the inequalities $|Y_n| \geqslant n$ occur and hence

$$\Pr(\lim_{n\to\infty} n^{-1} |\sum_{1}^{n} Y_j| = \infty) = 1.$$

It is instructive to compare the strong law Theorem 5.4.5 with the weak laws of § 4.2. Theorem 4.2.2 asserted that a weak law holds if Ee^{isY} is differentiable at the origin, showing that a first moment is not necessary for the convergence in probability of $\{n^{-1}S_n\}$. The complete statement of the weak law for identical summands is

Theorem 5.4.6 (Weak law of large numbers) Let

$$S_n = Y_1 + Y_2 + \ldots + Y_n,$$

where the Y_j are independently and identically distributed as Y. For constants $b_n, n = 1, 2, \ldots$, to exist such that

$$\lim_{n\to\infty} \Pr(|n^{-1}S_n - b_n| < \varepsilon) = 1$$

for every $\varepsilon > 0$, it is necessary and sufficient that

$$\lim_{t\to\infty} t \Pr(|Y| \geqslant t) = 0. \tag{5.4.4}$$

If (5.4.4) is true we may take

$$b_n = \int_{-n}^{n} y\, f(y)\, dy = ET_n,$$

where T_n is the truncated variable of (5.4.3). □

Part of the proof is outlined in Exercises 9–11.

From Lemma 1 it is necessary and sufficient for the strong law that $\Pr(|Y| \geqslant t)$ be integrable. If this is so the monotonicity of $\Pr(|Y| \geqslant t)$ implies (5.4.4) but the converse is not generally true [e.g. $\Pr(|Y| \geqslant t) \sim c(t \log t)^{-1}$]. The distinction between the two laws lies then in the tail behaviour of the distribution function of Y; the strong law requires the convergence to zero in (5.4.4) to be fast enough to ensure the integrability of the tail probabilities.

Exercises

1. The converse of Theorem 5.4.2 is not true and in general the almost sure convergence of $\sum X_n$ implies nothing about the convergence of sums of moments. Verify this assertion by considering the sequence $\{X_n\}$ of independent nonnegative variables with distributions

$$\Pr(X_n = 0) = 1 - n^{-2}, \qquad \Pr(X_n = n) = n^{-2}.$$

Show that for any $\varepsilon > 0$

$$\Pr\left(\sum_{j=n}^{\infty} X_j \geqslant \varepsilon\right) \leqslant \sum_{j=n}^{\infty} j^{-2}$$

and hence $\sum X_n < \infty$ (a.s.). However $\sum \operatorname{Var} X_n$ and indeed $\sum EX_n$ diverge. (The theorem has a converse if the X_n have zero means and are uniformly bounded.)

2. Show that if the X_n are independent then the probability is either 0 or 1 that $\sum X_n$ converges.

3. A truncation different from (5.4.3) is

$$T_n^{(b)} = \begin{cases} X_n & \text{if } |X_n| < b, \\ 0 & \text{if } |X_n| \geqslant b, \end{cases}$$

where b is some fixed positive number. A general condition for convergence of sums is *Kolmogorov's Three Series Theorem*: If $X_1, X_2, \ldots,$ are independent random variables a necessary and sufficient condition for

$$\Pr\left(\sum X_n \text{ converges}\right) = 1$$

is that each of the series $\sum \Pr(T_n^{(b)} \neq X_n)$, $\sum ET_n^{(b)}$, $\sum \operatorname{Var} T_n^{(b)}$ converge. The proof is not easy but find whether or not the theorem holds when

(a) $X_n \frown N(0, \sigma_n^2)$ and $\sum \sigma_n^2 < \infty$,
(b) $\Pr(X_n \leqslant x) = 1 - e^{-\lambda_n x}$, $x > 0$, and $\sum \lambda_n^{-1} < \infty$,
(c) $\Pr(X_n = 0) = 1 - 2n^{-\delta}$, $\Pr(X_n = \pm n) = n^{-\delta}$, $\delta > 0$.

4. Theorem 5.2.3 shows that the almost sure convergence of $\sum X_n$ implies that $\sum X_n$ converges in probability. The converse is true when the summands are independent, (see Exercise 5.2.7). Verify the assertion for the relevant sequences in Exercise 3.

5. Lemma 1 can be restated as $E|X| < \infty$ if and only if

$$P(\limsup_{n \to \infty} \{\omega \,|\, |X_n| \geqslant n\}) = 0.$$

6. The classical laws of large numbers and central limit theorem concern independent identically distributed variables. For more general sequences various possibilities exist. Instances of widely differing behaviour can be constructed by considering independent variables $X_1, X_2, \ldots$, with distributions

$$\Pr(X_n = \pm b_n) = p_n, \qquad \Pr(X_n = 0) = 1 - 2p_n,$$

and choosing b_n and p_n appropriately. Give examples. In particular show that it is possible for the central limit theorem to hold but not the laws of large numbers.
Ref: FELLER I, Chapter 10. Also see Exercise 4.2.9.

7. Since $\sum_{j=2}^{\infty} (j^2 \log |j|)^{-1} = 2c^{-1}$ is convergent the numbers

$$p_j = \frac{c}{j^2 \log |j|}, \qquad j = \pm 2, \pm 3, \ldots,$$

form a symmetric probability distribution. Verify that

$$\psi(s) = \sum_{|j| \geqslant 2} e^{isj} p_j$$

has a derivative which is the sum of a series uniformly convergent in an interval containing the origin. Consequently $\psi'(0)$ exists and

if $p_y = \Pr(Y = y)$ the weak law holds. Since $E|Y| = \infty$, the strong law does not.

8. State which values of k ensure that one (or both) of the laws of large numbers apply when

$$\Pr(Y = y) = \frac{c(k)}{j^2(\log|j|)^k}, \quad j = \pm 2, \pm 3, \ldots$$
$$= 0 \quad \text{otherwise.}$$

A proof of Theorem 5.4.6 can be found in FELLER II, page 232. However, sufficiency can be established by the following, which parallel the proof for the strong law.

9. Use Chebyshev's inequality and a method similar to that of Lemma 2 to prove for the truncated variable $\{T_n\}$ of (5.4.3) that if

$$\lim_{n\to\infty} n^{-1} \int_{-n}^{n} y^2 f(y)\, dy = 0$$

then for every $\varepsilon > 0$

$$\lim_{n\to\infty} \Pr\left(n^{-1}\left|\sum_{j=1}^{n} (T_j - ET_j)\right| \geqslant \varepsilon\right) = 0.$$

10. Next, show that (5.4.4) implies $\lim_{t\to\infty} t^{-1} \int_{-t}^{t} y^2 f(y)\, dy = 0$.

11. A further consequence of (5.4.4) is that if $n^{-1}\sum_{j=1}^{n} T_j - b_n \xrightarrow{P} 0$ for some constants b_n, then $n^{-1}S_n - b_n \to 0$ also.

Combining these three results shows that (5.4.4) is sufficient for the weak law.

5.5* OCCUPATION TIMES AND RECURRENT MARKOV CHAINS

Let $\{X_n, n = 0, 1, 2, \ldots,\}$ be a homogeneous, irreducible Markov chain on the non-negative integers (or a subset thereof) and which, for convenience, we assume to be aperiodic. As in § 3.5, § 4.5 write

* To be omitted at a first reading.

the transition and first passage probabilities respectively as

$$p_{ij}^{(n)} = \Pr(X_n = j \mid X_0 = i)$$
$$f_{ij}^{(n)} = \Pr(X_n = j;\, X_\nu \neq j,\, \nu = 1,2,\ldots,n-1 \mid X_0 = i).$$

According to Definitions 4.5.1, 2, state j is recurrent if and only if $\sum_{n=0}^{\infty} p_{jj}^{(n)} = \infty$, or equivalently $F_{jj} = \sum_{n=1}^{\infty} f_{jj}^{(n)} = 1$, and Theorem 4.5.4 asserts that all states of an irreducible chain are positive recurrent, null recurrent, or transient together.

In the context of weak convergence the important distinction is between positive recurrence as opposed to null recurrence or transience (Theorems 4.5.6, 7), as it is only in the former case that the transition probabilities converge to a limiting distribution $\{\pi_j\}$, $\pi_j = \lim_{n\to\infty} p_{ij}^{(n)}$. On the other hand the use of strong convergence leads to a characterisation in probabilistic terms which shows that the behaviour of $\{X_n\}$ when recurrent is radically different to its evolution through time when transient. Furthermore a version of the strong law of large numbers for dependent variables leads to an interpretation of the limits π_j in the positive recurrent case (Theorem 5.5.3).

To motivate the discussion, note that the classical strong law of large numbers imputes a certain repetitive behaviour to the sequence of averages $\{n^{-1}S_n\}$ in the sense that the probability is one that infinitely many $n^{-1}S_n$ take values in (or visit) an arbitrarily small neighbourhood about the mean μ. The distinction we wish to make between recurrent and transient irreducible chains is of the same nature, namely that recurrence is equivalent to being certain that each state is visited infinitely often as the chain evolves with time, whereas the probability of infinitely many visits from one state to another is zero if the chain is transient. Recurrent chains therefore possess a repetitive quality that transient chains lack.

The formal statement and proof of this result is conveniently phrased in terms of occupation times or number of visits between pairs of states. Given $X_0 = i$ and j any other state, let

$$I_{ij}^{(\nu)} = \begin{cases} 1 & \text{if } X_\nu = j, \\ 0 & \text{otherwise.} \end{cases}$$

Starting from i, the number of times j is visited in the time interval $[1,n]$ is the *occupation time of j in $[1,n]$* defined by

$$N_{ij}^{(n)} = \sum_{\nu=1}^{n} I_{ij}^{(\nu)}. \tag{5.5.1}$$

Since $EI_{ij}^{(\nu)} = \Pr(X_\nu = j \mid X_0 = i) = p_{ij}^{(\nu)}$ we have

$$EN_{ij}^{(n)} = \sum_{\nu=1}^{n} p_{ij}^{(\nu)}.$$

For $n = 1,2,\ldots$, the random variable $N_{ij}^{(n)}$ are non-negative and non-decreasing. Hence

$$N_{ij} = \sum_{\nu=1}^{\infty} I_{ij}^{(\nu)}$$

is well defined although it may take the value $+\infty$. N_{ij} is the occupation time of j, given initial state i, over an arbitrarily long period of time, and its expectation is

$$EN_{ij} = \sum_{\nu=1}^{\infty} p_{ij}^{(\nu)}.$$

From Definitions 4.5.1, 4.5.2 and Theorem 4.5.4 we immediately have a criterion for recurrence in terms of occupation times:

Theorem 5.5.1 An irreducible chain is recurrent if and only if there is a state i for which $EN_{ij} = \infty$. □

Our aim is to establish that $N_{ij} = \sum_{\nu=1}^{\infty} I_{ij}^{(\nu)}$ diverges (a.s.) to $+\infty$ (i.e. the probability is 1 that j is visited from i infinitely often) if and only if $\{X_n\}$ is recurrent. A consequence of Theorem 5.5.1 is that recurrence is equivalent to $EN_{ij} = \infty$, all i,j, but this in itself does not yield the desired result. Indeed we cannot appeal directly to any of the theorems of § 5.4. since the $I_{ij}^{(\nu)}$, $\nu = 1,2,\ldots$, are not independent (see Exercise 9). However a simple proof which uses the Markov property of $\{X_n\}$ in an essential way can be constructed.

Theorem 5.5.2 If i and j are any two states of the irreducible Markov chain $\{X_n\}$ the probability is one that, starting from i, j is visited infinitely often if and only if the chain is recurrent and is zero otherwise.

Proof. For each fixed pair (i,j) of states the probability that, starting at i, there are at least m visits to j in $[1, \infty)$ is

$$g_{ij}(m) = \Pr(N_{ij} \geqslant m).$$

This quantity is monotonic in m and the limit

$$g_{ij} = \lim_{m \to \infty} g_{ij}(m) = \Pr\Big(\sum_{\nu=1}^{\infty} I_{ij}^{(\nu)} = \infty\Big)$$

is the probability that j is visited infinitely often from i. We show that recurrence is equivalent to $g_{ij} = 1$ and transience to $g_{ij} = 0$.

Since the probability of remaining forever in the initial state is

$$\lim_{n \to \infty} (p_{ii})^n = 0,$$

at least m visits to j occur if and only if there is a first visit to j at time ν say, and subsequently at least $m-1$ returns to j. Using the Markov property it follows that for $m = 2, 3, \ldots,$

$$g_{ij}(m) = \sum_{\nu=1}^{\infty} f_{ij}^{(\nu)} g_{jj}(m-1) = F_{ij} g_{jj}(m-1),$$

and (5.5.2)

$$g_{ij}(1) = F_{ij}.$$

When $i = j$ these equations reduce to a recursion formula in m whose solution is easily found to be

$$g_{jj}(m) = (F_{jj})^m.$$

Hence

$$g_{ij}(m) = F_{ij}(F_{jj})^{m-1}.$$

But $F_{ij} = 1$ if and only if the chain is recurrent, in which case $g_{ij}(m) = 1$ identically. Since $F_{jj} \leqslant 1$, the only other possibility is

$$\lim_{m \to \infty} g_{ij}(m) = 0$$

which therefore corresponds to transience. □

Note that the theorem holds whether or not the chain is aperiodic.

If $\{X_n\}$ is recurrent, so that $N_{ij}^{(n)} \overset{\text{a.s.}}{\to} \infty$, it is natural to enquire whether the average $n^{-1} N_{ij}^{(n)}$ converges to a limit. For aperiodic chains

$$\lim_{n \to \infty} p_{ij}^{(n)} = \pi_j$$

exists for all i, j (Theorems 4.5.1, 4.5.3)* and hence as $n \to \infty$

$$E(n^{-1} N_{ij}^{(n)}) = n^{-1} \sum_{\nu=1}^{n} p_{ij}^{(\nu)} \to \pi_j.$$

We anticipate a strong law of large numbers

$$n^{-1} N_{ij}^{(n)} \xrightarrow{\text{a.s.}} \pi_j. \tag{5.5.3}$$

for the sequence $\{I_{ij}^{(\nu)}\}$ but, as with the previous theorem, this cannot be directly deduced from earlier results since the $I_{ij}^{(\nu)}$ are not independent.

Attention is restricted to positive recurrent chains, and the argument depends on the introduction of a sequence of independent variables which are related to the $N_{ij}^{(n)}$ in such a way that (5.5.3) can be deduced from the strong law Theorem 5.4.5. These random variables turn out to be the *first passage times* T_{ij} induced by the first passage distributions, thus

$$f_{ij}^{(n)} = \Pr(T_{ij} = n), \qquad n = 1, 2, \ldots$$

(See Exercise 3.5.7 for a special case). T_{ij} is the time taken for the chain to reach j for the first time from i. For recurrent irreducible chains.

$$\Pr(T_{ij} < \infty) = F_{ij} = 1$$

for all (i,j), and, from Theorem 4.5.7,

$$ET_{jj} = \mu_{jj} = 1/\pi_j < \infty \tag{5.5.4}$$

in the positive recurrent case.

Let $T_{jj}^{(1)}, T_{jj}^{(2)}, \ldots,$ be independent and identically distributed as T_{jj}. A consequence of the Markov property is that, starting from i, the time of the rth visit to j is the sum of the independent variables $T_{ij}, T_{jj}^{(1)}, \ldots, T_{jj}^{(r-1)}$. Hence $N_{ij}^{(n)} = r$ if and only if the sum of these r passage times is $\leqslant n$ simultaneously with the sum of $r+1$ passage times being $> n$. More precisely,

$$\{\omega \mid N_{ij}^{(n)} = r\} = \{\omega \mid T_{ij} + \sum_{\nu=1}^{r-1} T_{jj}^{(\nu)} \leqslant n < T_{ij} + \sum_{\nu=1}^{r} T_{jj}^{(\nu)}\},$$

* When periodic $\lim_{n\to\infty} n^{-1} \sum_{v=1}^{n} p_{ij}^{(v)} = \pi_j$ also but for simplicity our discussion is limited to the aperiodic case.

which is conveniently reformulated as

$$N_{ij}^{(n)} \geqslant r \text{ if and only if } T_{ij} + \sum_{\nu=1}^{r-1} T_{ij}^{(\nu)} \leqslant n. \tag{5.5.5}$$

Since the addition of a finite number of terms does not affect the limit value of an average, the strong law Theorem 5.4.5 ensures that for positive recurrent chains

$$n^{-1}(T_{ij} + \sum_{\nu=1}^{n-1} T_{jj}^{(\nu)}) \overset{\text{a.s.}}{\to} \mu_{jj}. \tag{5.5.6}$$

We show that (5.5.6) taken in conjunction with (5.5.4) and (5.5.5) implies (5.5.3).

Theorem 5.5.3 If the aperiodic irreducible Markov chain $\{X_n\}$ is positive recurrent then for each pair (i,j),

$$\Pr(\lim_{n\to\infty} n^{-1} N_{ij}^{(n)} = \pi_j \mid X_0 = i) = 1,$$

where $\{\pi_j\}$ is the limit distribution given by $\pi_j = \lim_{n\to\infty} p_{ij}^{(n)}$.

Proof. A simple restatement is that we want to show that the hypothesis implies that for $\varepsilon > 0$, the probability is one that infinitely many of the inequalities

$$|n^{-1} N_{ij}^{(n)} - \pi_j| \leqslant \varepsilon, \qquad n = 1, 2, \ldots, \tag{5.5.7}$$

hold. From (5.5.6), with probability one,

$$|n^{-1}(T_{ij} + \sum_{\nu=1}^{n-1} T_{jj}^{(\nu)}) - \mu_{jj}| \leqslant \delta \tag{5.5.8}$$

is true for infinitely many n, where $\delta > 0$ is arbitrary. The theorem is proved if it can be shown that (5.5.8) implies (5.5.7) with ε decreasing to zero as δ does.

Firstly note that, from (5.5.5),

$$T_{ij} + \sum_{\nu=1}^{n-1} T_{jj}^{(\nu)} \leqslant n(\mu_{jj} + \delta) \tag{5.5.9}$$

implies, with $[n(\mu_{jj}+\delta)]$ the integer part of $n(\mu_{jj}+\delta)$,

$$\begin{aligned}\frac{N_{ij}^{[n(\mu_{jj}+\delta)]}}{[n(\mu_{jj}+\delta)]}-\pi_j &\geqslant \frac{n}{[n(\mu_{jj}+\delta)]}-\pi_j \\ &\geqslant \frac{n}{n(\mu_{jj}+\delta)}-\frac{1}{\mu_{jj}} \\ &= \frac{-\delta}{\mu_{jj}(\mu_{jj}+\delta)}.\end{aligned}$$

Therefore

$$n^{-1}N_{ij}^{(n)}-\pi_j \geqslant -\varepsilon$$

holds whenever (5.5.9) is true, and $\varepsilon = c\delta[\mu_{jj}(\mu_{jj}+\delta)]^{-1}$, with c some positive constant, tends to zero with δ. By a similar argument

$$T_{ij}+\sum_{\nu=1}^{n-1} T_{jj}^{(\nu)} \geqslant n(\mu_{jj}-\delta) \qquad (5.5.10)$$

implies

$$\begin{aligned}\frac{N_{ij}^{[n(\mu_{jj}-\delta)]+1}}{[n(\mu_{jj}-\delta)]+1}-\pi_j &\leqslant \frac{n}{[n(\mu_{jj}-\delta)]+1}-\pi_j \\ &\leqslant \frac{1}{\mu_{jj}-\varepsilon}-\frac{1}{\mu_{jj}}=\frac{\delta}{\mu_{jj}(\mu_{jj}-\delta)},\end{aligned}$$

or equivalently,

$$n^{-1}N_{ij}^{(n)}-\pi_j \leqslant \varepsilon$$

for n sufficiently large. But (5.5.8) is the same as (5.5.9) and (5.5.10) being true simultaneously. Hence (5.5.8) implies (5.5.7) which it was required to prove. □

The theorem exhibits an important example of a sequence of dependent variables obeying the strong law, a situation by no means general. Furthermore we can now attach a probabilistic meaning to the π_j as the almost sure limits of the average occupation times $n^{-1}N_{ij}^{(n)}$, i.e. we have a frequency interpretation of the stationary distribution.

Finally, consider some relevant aspects of the special Markov chain $\{S_n\}$, with $S_n = Y_1+Y_2+\ldots+Y_n$, the nth partial sum of independent integer valued variables identically distributed as Y.

Recall from Example 3.5.2 that the transition probabilities are

$$p_{ij} = \Pr(Y = j-i),$$

and (5.5.11)

$$p_{ij}^{(n)} = \Pr(S_n = j \mid S_0 = i) = \Pr(S_n = j-i).$$

In Example 3.5.4 this chain was referred to as unrestricted random walk. It differs from the sequence considered so far in this section in the sense that it is possible for $\{S_n\}$ to drift either to $+\infty$ or $-\infty$ as n increases, whereas chains on the non-negative half line diverge to $+\infty$ if they diverge at all. A consequence is that $\{S_n\}$ can only be null recurrent or transient and the possibility of a limit distribution $\{\pi_j\}$ does not arise for unrestricted walk*. The next theorem, which is an extension of Exercise 4.5.1, makes this clear for those walks in which the increment variable Y has finite mean and variance.

Theorem 5.5.4 Unrestricted walk, with transition probabilities (5.5.11), and $\operatorname{Var} Y = \sigma^2 < \infty$ is recurrent if and only if $EY = \mu = 0$. Such a chain can never be positive recurrent.

Proof. By the central limit theorem for lattice variables (Theorem 4.3.3)

$$\Pr(S_n = 0) \sim \frac{1}{\sigma(2\pi n)^{\frac{1}{2}}} e^{-n\mu^2/(2\sigma^2)} \qquad \text{as } n \to \infty. \qquad (5.5.12)$$

Therefore the series

$$\sum_{n=1}^{\infty} p_{ii}^{(n)} = \sum_{n=1}^{\infty} \Pr(S_n = 0) \quad \text{and} \quad \sum_{n=1}^{\infty} (\sigma^2 2\pi n)^{-\frac{1}{2}} \exp\{-n\mu^2/(2\sigma^2)\}$$

converge or diverge together. Hence $\sum p_{ii}^{(n)} = \infty$ and if and only if $\mu = 0$. That positive recurrence is impossible when $\mu = 0$ follows also from (5.5.12) with $\mu = 0$. □

The theorem continues to hold without the assumption of finite variance but a more sophisticated proof is required. A curious byproduct is that for walks with $\mu = 0$ the probability is one that the starting state is revisited infinitely often but the times between each visit have infinite expectation.

* For simple random walk this was shown in Exercise 4.5.2.

s

Exercises

1. Deduce (5.5.2) from (5.5.5).

2. A consequence of (5.5.5) is

$$\Pr(N_{ij}^{(n)} = r) = \Pr(T_{ij} + \sum_{\nu=1}^{r-1} T_{jj}^{(\nu)} \leqslant n) - \Pr(T_{ij} + \sum_{\nu=1}^{r} T_{jj}^{(\nu)} \leqslant n), \quad r \geqslant 1,$$

$$\Pr(N_{ij}^{(n)} = 0) = 1 - \Pr(T_{ij} \leqslant n),$$

and hence (with $F_{ij}(z) = \sum_{n=1}^{\infty} z^n f_{ij}^{(n)}$),

$$\sum_{n=1}^{\infty} z^n \Pr(N_{ij}^{(n)} = r) = \frac{F_{ij}(z)\{1-F_{jj}(z)\}}{1-z}\{F_{jj}(z)\}^{r-1}, \quad r \geqslant 1.$$

The generating functions of the occupation time distributions are therefore given by

$$\sum_{n=1}^{\infty} z^n E x^{N_{ij}^{(n)}} = \frac{1-F_{ij}(z)+x\{F_{ij}(z)-F_{jj}(z)\}}{(1-z)\{1-xF_{jj}(z)\}}$$

3. (Continuation.) Generating functions for moments:

$$\sum_{n=1}^{\infty} z^n E N_{ij}^{(n)} = \frac{F_{ij}(z)}{(1-z)\{1-F_{jj}(z)\}}$$

$$\sum_{n=1}^{\infty} z^n E(N_{ij}^{(n)})^2 = \frac{F_{ij}(z)\{1+F_{jj}(z)\}}{(1-z)\{1-F_{jj}(z)\}^2}.$$

4. (Continuation.) If the state space of $\{X_n\}$ is $\{0,1\}$ show that the first passage distributions are geometric and find expressions for the first two moments of $N_{oo}^{(n)}$.

5. For a transient chain the total occupation time N_{ij} is a proper random variable whose distribution is modified geometric

$$\Pr(N_{ij} = 0) = 1 - F_{ij}$$
$$\Pr(N_{ij} = r) = F_{ij}(1-F_{jj})F_{jj}^{r-1}, \qquad r = 1, 2, \ldots,$$

where $F_{ij} = \sum_{n=1}^{\infty} f_{ij}^{(n)}$. Write down these probabilities for simple random walk on the non-negative integers (Example 4.5.1).

6. Whether or not $\{X_n\}$ is recurrent, show that

$$\lim_{n\to\infty} \frac{EN_{ij}^{(n)}}{1+EN_{jj}^{(n)}} = F_{ij}.$$

[*Hint*: The transient and recurrent cases require separate proof and for the latter use the result stated in Exercise 2.2.11].

7. Suppose $\{X_n\}$ is positive recurrent with stationary distribution $\{\pi_j\}$ and Var $T_{jj} = \sigma_{jj}^2 < \infty$. The central limit theorem holds for the sequence $T_{ij}, T_{jj}^{(1)}, T_{jj}^{(2)}, \ldots$ Use (5.5.5) to show that

$$\lim_{n\to\infty} \Pr\left(\frac{N_{ij}^{(n)} - n\pi_j}{\sigma_{jj}\pi_j^{3/2}n^{\frac{1}{2}}} \leqslant x\right) = \frac{1}{(2\pi)^{\frac{1}{2}}} \int_{-\infty}^{x} e^{-\frac{1}{2}y^2}\,dy.$$

8. For simple random walk on the non-negative half-line (Example 4.5.1), use Exercise 7 to find an approximation for the probability that, given $X_0 = 0$, the number of returns to the origin lies between 40 and 60 when $n = 75$ and $p = \frac{1}{4}$.

9. Verify that the entries in the table below are the joint probabilities $\Pr(I_{ij}^{(n)} = k, I_{ij}^{(n+m)} = r)$:

		$I_{ij}^{(n+m)}$ 0	1
$I_{ij}^{(n)}$	0	$1 - p_{ij}^{(n+m)} - p_{ij}^{(n)}(1 - p_{jj}^{(m)})$	$p_{ij}^{(n+m)} - p_{ij}^{(n)}p_{jj}^{(m)}$
	1	$p_{ij}^{(n)}(1 - p_{jj}^{(m)})$	$p_{ij}^{(n)}p_{jj}^{(m)}$

Clearly $I_{ij}^{(n)}$ and $I_{ij}^{(n+m)}$ are not independent and in particular

$$\text{Cov}(I_{ij}^{(n)}, I_{ij}^{(n+m)}) = p_{ij}^{(n)}(p_{jj}^{(m)} - p_{ij}^{(n+m)}).$$

10. (Continuation.) Given that state j was visited at time n, the probability of a visit at time $n+m$ is $p_{jj}^{(m)}$, independent of n. On the other

hand

$$\Pr(X_{n+m} = j \mid X_n \neq j) = \Pr(I_{ij}^{(n+m)} = 1 \mid I_{ij}^{(n)} = 0)$$
$$= (p_{ij}^{(n+m)} - p_{ij}^{(n)} p_{jj}^{(m)})/(1 - p_{ij}^{(n)}).$$

Extend these results by considering the joint distribution of $I_{ij}^{(n)}$, $I_{ij}^{(n+m)}$, $I_{ij}^{(n+m+v)}$, and explain how the observation of visit to j at n and $n+m$ can be used to predict whether or not a visit will occur at time $n+m+v$.

ANSWERS TO SELECTED EXERCISES

§ 1.2, page 13

9. (a) $F(x_5)-F(x_1)+F(x_{-1})-F(x_{-5})$,
   (b) $F(x_{n+1})-F(x_1)+F(x_{-1})-F(x_{-n-1})$,
   (c) $F(x_3)-F(x_2)+F(x_1)-F(x_{-1})+F(x_{-2})-F(x_{-3})$,
   (d) $F(\lim_{n\to\infty} x_n)-F(\lim_{n\to\infty} x_{-n})$,
   (e) $F(x_1)-F(x_{-1})$.
12. (a) $\frac{1}{4}$, (b) $P(A \cup B) = \frac{5}{8}$, $P(\bar{A}\bar{B}) = \frac{3}{8}$.
16. $\Omega = \{\omega \mid \omega = 1, 2, 3, \ldots\}$, (a) $1-M/N$,
   (b) $(1-M/N)^2\{1-(1-M/N)^4\}$.

§ 1.3, page 22

3. The required probabilities for Question 1 are 1/6, 1/6, 1/6, 1/36, and for Question 2, 11/36, 1/2, 5/18, 1/6, respectively.
5. (i) 2/5, (ii) 3/10, (iii) 3/10, (iv) 9/10.
6. (i) 13/25, (ii) 9/25, (iii) 6/25, (iv) 21/25.
8. (a) 1/2, (b) 1/8.

§ 1.4, page 32

2. $\Pr(X = j) = (j-1)/36, j = 2, 3, \ldots, 7;$
   $\Pr(X = j) = (13-j)/36, j = 8, 9, \ldots, 12.$
7. $1-e^{-3}, 1-e^{-2}, e^{-1}-e^{-3}, e^{-1}-e^{-2}, 1-e^{-1}$.
9. $p = \frac{1}{2}$.
11. $\Pr(\log X \leqslant x) = 1-\exp(-\lambda e^x)$.

§ 1.5, page 44

2. $EY = p+4q$, $\operatorname{Var} Y = 9pq$.
6. $\Pr(|X-\mu| \geqslant 2\sigma) \leqslant 0{\cdot}25$. Exact results (a) 0·046, (b) 0·050.
8. $EX^k < \infty$ if and only if $\alpha > k$.
11. Expected gain $= -\$\,0{\cdot}25$.

§ 2.2, page 70

4. $G(z^3)$, $zG(z)$, $z^bG(z^a)$.
6. If $G(z) = \sum_{n=1}^{\infty} z^n \Pr(X = n)$ then

$$\sum_{n=1}^{\infty} z^n \Pr(X = n+1) = \{G(z) - \Pr(X = 1)\}/z$$

and

$$\sum_{n=1}^{\infty} z^n \Pr(X < n) = G(z)(1-z)^{-1}.$$

7. $R(z) = \{\mu - Q(z)\}/(1-z)$ and $R(1) = \frac{1}{2}EX(X-1)$.

§ 2.3, page 86

4. (a) $(2\sigma\sqrt{\pi})^{-1}e^{-x^2/(4\sigma^2)}$ and $e^{-\sigma^2}$.

   (b) $\frac{1}{c}\left(1 - \frac{|x|}{c}\right), |x| < c$, and $\frac{2[1 - \cos cs]}{s^2c^2}$

6. $c = 2\mu/\mu_2$. If $\mu_j = EX^j$ the required mean and variance are $\frac{\mu_3}{3\mu_2}$ and $\frac{\mu_4}{6\mu_2} - \left(\frac{\mu_3}{3\mu_2}\right)^2$ respectively.

8. $\{\int_0^\infty xf(x)\,dx\}^{-1}$.
9. All positive c_1, c_2 such that $c_1 - 2c_1c_2 + c_2 = 0$.

16. $1 + is3 + \frac{(is)^2}{2!}\left(\frac{21}{2}\right) + o(s^2)$.

§ 2.4, page 96

1. If $f_X(x) = (2\pi)^{-1}, -\pi < x < \pi$, then $f_y(y)$ is
   (a) $\pi^{-1}, 0 < y < \pi$,
   (b) $\{2\pi\sqrt{(ay)}\}^{-1}, 0 < y < a\pi^2$,
   (c) $\{\pi(1+y^2)\}^{-1}, -\infty < y < \infty$.

§ 2.5, page 105

3. The gamma density (2.5.5) has mean and variance both equal to α and $EX^r = \alpha(\alpha+1)\ldots(\alpha+r-1)$. The rth moment EX^r of the beta density (2.5.9) is $\Gamma(\alpha+\beta)\Gamma(\alpha+r)\{\Gamma(\alpha+\beta+r)\Gamma(\alpha)\}^{-1}$ with mean $\alpha(\alpha+\beta)^{-1}$ and variance $(\alpha\beta)\{(\alpha+\beta)^2(\alpha+\beta+1)\}^{-1}$.
7. (a) $N^n/n!$, (b) $\{\lambda^n(1-\lambda)^{N-n}[2\pi n(1-\lambda)]^{\frac{1}{2}}\}^{-1}$.
9. Writing $p = M/N$ then

$$EX = np, \operatorname{Var} X = np(1-p)\{1-(n-1)/(N-1)\}.$$

The corresponding binomial moments are np and $np(1-p)$.

11. $$EX^4 = \frac{\{11-3(n+M)+nM\}}{N-3}\{EX^3-3EX^2+2EX\}$$
$$+7EX^3-14EX^2-8EX.$$

12. (a) $\{(y, x]\}$, (b) $\{(-\infty, \infty)\}$, (c) $\{(-\infty, x]\}$, (d) $\{(-\infty, y]\} \cup \{(x, \infty)\}$.

§ 3.1, page 119

1. (a) 1/18, (b) 1/9, (c) 0, (d) 1/18.
4. Each of $\Pr(X = -1, Y = 1)$, $\Pr(X = 1, Y = 1)$, $\Pr(X = -2, Y = 4)$, $\Pr(X = 2, Y = 4)$ are equal to $\frac{1}{4}$.
8. $\mu_X = -1$, $\mu_Y = 2$, $\sigma_X^2 = 3/10$, $\sigma_Y^2 = 9/5$, $\rho = 1/\sqrt{6}$, and $c = \sqrt{5/(3\pi)}$.
10. $f_X(x) = e^{-x}$, $x > 0$; $f_Y(y) = ye^{-y}$, $y > 0$; $E\exp(-s_1X - s_2Y) = \{(s_2+1)(s_1+s_2+1)\}^{-1}$.

§ 3.2, page 128

4. (a) 1/2, (b) 1/2, (c) 1/3.
5. $(1-p)^2$, where p is the probability of a head.
6. (a) $\binom{36}{10}\Big/\binom{39}{13}$, (b) $\binom{36}{13}\Big/\binom{39}{13}$.
8. Any choice of the a_{ij} such that

$$a_{11}a_{21}\operatorname{Var} X_1 + a_{12}a_{22}\operatorname{Var} X_2 + (a_{11}a_{22}+a_{12}a_{21})\operatorname{Cov}(X_1, X_2) = 0,$$

e.g. $a_{12} = 0$ and $a_{21} = -a_{22}\operatorname{Cov}(X_1, X_2)/\operatorname{Var} X_1$.

15. (a) $$\Pr(X(t_1+t_2) = k) = \exp(-\lambda(t_1+t_2))\frac{[\lambda(t_1+t_2)]^k}{k!}$$

(b) $$\Pr(X(t_1) = j \mid X(t_1+t_2) = k) = \binom{k}{j}\frac{t_1^j t_2^{k-j}}{(t_1+t_2)^k}, j = 0, 1, \ldots, k.$$

16. $Ee^{-sZ} = \exp\{-s\mu_x + \frac{1}{2}s^2\rho^2\sigma_x^2\}$, $EZ = \mu_x$, $\operatorname{Var} Z = \rho^2\sigma_x^2$.

§3.3, page 139

3. $\Pr(Y \leqslant x) = [F(x)]^n$, $\Pr(Z \leqslant x) = 1-[1-F(x)]^n$.
4. (a) $3p(1-p)^3$, (b) $p(1-p)^3$, (c) $3p^2(1-p)+p^3$.

7. The probability of winning j times is $\binom{n}{j}\left(\frac{2}{5}\right)^j\left(\frac{3}{5}\right)^{n-j}$, $j=0, 1, \ldots, n$, with mean $2n/5$ and variance $6n/25$. The probability of a gain of \$$k$ is $\binom{n}{(k+2n)/5}\left(\frac{2}{5}\right)^{(k+2n)/5}\left(\frac{3}{5}\right)^{(3n-k)/5}$, $k=-2n, 5-2n, \ldots, 3n$.

12. (a) $\Pr(S_n = j) = \binom{Nn}{j} p^j(1-p)^{Nn-j}, \quad j = 0, 1, \ldots, Nn$

(b) $\Pr(S_n = j) = e^{-\lambda n}\dfrac{(\lambda n)^j}{j!}, \quad j = 0, 1, \ldots,$

(c) $\Pr(S_n = j) = \binom{-n}{j}(-1)^j\rho^j(1-\rho)^n, j = 0, 1, \ldots,$

(d) $f_{s_n}(x) = \dfrac{e^{-\lambda x}\lambda^n x^{n-1}}{(n-1)!}, \quad 0 < x.$

14. (a) $f_y(y) = (2\sigma\sqrt{\pi})^{-1}\exp(-y^2/4\sigma^2),$
$-\infty < y < \infty, Ee^{isY} = e^{-s^2\sigma^2}.$

(b) $f_y(y) = (\lambda/2)e^{-|\lambda y|}, -\infty < y < \infty, Ee^{isY} = \lambda^2/(\lambda^2 + s^2).$

§ 3.4, page 146

1. Apply (3.4.3) with B = 'drawn ball is white' and A_j = 'j white balls in first move', $j = 0, 1, 2, 3, 4$. Then $P(A_j) = \binom{4}{j}\binom{4}{4-j}\Big/\binom{8}{4}$ and the required probability is $P(A_4|B) = 3/97$.

6. With likelihood $\Pr(Y = y|p) = \binom{n}{y}p^y(1-p)^{n-y}, y = 0, 1, \ldots, n,$ and prior density $f(p) = [B(a, b)]^{-1} p^{a-1}(1-p)^{b-1}$, the posterior density is $f(p|y) = [B(a+y, b+n-y)]^{-1} p^{a+y-1}(1-p)^{b+n-y-1}$. The prior and posterior means are respectively $a/(a+b)$, $(a+y)/(a+b+n)$. The maximum frequencies are

$$\frac{(a-1)^{a-1}(b-1)^{b-1}}{B(a,b)(a+b-2)^{a+b-2}}$$

and

$$\frac{(a+y-1)^{a+y-1}(b+n-y-1)^{b+n-y-1}}{B(a+y,b+n-y)(a+b+n-2)^{a+b+n-2}}$$

§ 3.5, page 154

2. $G_{00}(z)$ and $G_{01}(z)$ satisfy $(1 - zp_1)\,G_{00}(z) - zp_2G_{01}(z) = 1$ and $(1 - zq_2)G_{01}(z) = zq_1G_{00}(z)$. Hence

$$p_{00}^{(n)} = \{p_2 + q_1(p_1 - p_2)^n\}/(q_1 + p_2),$$
$$p_{01}^{(n)} = q_1\{1 - (p_1 - p_2)^n\}/(q_1 + p_2).$$

§ 4.1, page 166

2. Binomial 0·0338, Poisson 0·0404.
6. 0·7942.
7. Normal 0·917, Poisson 0·922.
8. 42.

§ 4.2, page 176

2. $n^{-1} \sum_{j=1}^{n} X_j^2 \xrightarrow{\text{P}} EX^2.$

§ 4.3, page 184

2. 0·9684.

§ 4.4, page 194

3. $f(z) = (1+z)^{-2}$.
6. Confidence interval is (0·85, 7·15) when $n = 6$ and (2·88, 5·12) when $n = 30$.

§ 4.5, page 208

6. $\pi_j = \dfrac{[1 - p/q](p/q)^j}{1 - (p/q)^K}.$
8. Period = 2.
10. $\lim_{n \to \infty} p_{00}^{(n)} = \frac{1}{2}$.
14. $\lim_{n \to \infty} p_{ij}^{(n)} = 0$ if $\sum_{\nu=1}^{\infty} f^{(\nu)} < 1$ or $\mu = \sum_{\nu=1}^{\infty} \nu f^{(\nu)} \doteq \infty$. Otherwise

$$p_{1j}^{(n)} \to \frac{1}{\mu} \sum_{\nu=j}^{\infty} f^{(\nu)}.$$

§ 5.1, page 220

2. 1/5, 2/5, 3/5, 1.
3. $p \geqslant 1/k$.

§ 5.2, page 227

1. $\Pr({}^*X = 0) = P(\{-1\}) + P(\{-2\})$, $\Pr({}^*X = 1) = P(\{1\})$, $\Pr({}^*X = 2) = P(\{2\})$, $\Pr({}_*X = -2) = P(\{-2\})$, $\Pr({}_*X = -1) = P(\{-1\})$, $\Pr({}_*X = 0) = P(\{1\}) + P(\{2\})$.

§ 5.3, page 234

1. In the notation of the lemma choose s_o and s_1 to minimise the right hand sides of $\Pr(S_n \leqslant n(\lambda^{-1} - \varepsilon) \leqslant (Ee^{-s_0 Z})^n$ and $\Pr(S_n \geqslant n(\lambda^{-1} + \varepsilon)) \leqslant (Ee^{-s_1 W})$. In the normal case $\Pr(|n^{-1}S_n - \mu| \geqslant \varepsilon) \leqslant 2\exp\{-n\varepsilon^2/(2\sigma^2)\}$.

INDEX